T0185959

SpringerBriefs in Applied Sciences and Technology

Computational Intelligence

Series Editor

Janusz Kacprzyk, Systems Research Institute, Polish Academy of Sciences, Warsaw, Poland

SpringerBriefs in Computational Intelligence are a series of slim high-quality publications encompassing the entire spectrum of Computational Intelligence. Featuring compact volumes of 50 to 125 pages (approximately 20,000–45,000 words), Briefs are shorter than a conventional book but longer than a journal article. Thus Briefs serve as timely, concise tools for students, researchers, and professionals.

More information about this subseries at http://www.springer.com/series/10618

Fevrier Valdez · Cinthia Peraza · Oscar Castillo

General Type-2 Fuzzy Logic in Dynamic Parameter Adaptation for the Harmony Search Algorithm

 Springer

Fevrier Valdez
Division of Graduate Studies
Tijuana Institute of Technology
Tijuana, Baja California, Mexico

Cinthia Peraza
Division of Graduate Studies
Tijuana Institute of Technology
Tijuana, Baja California, Mexico

Oscar Castillo
Division of Graduate Studies
Tijuana Institute of Technology
Tijuana, Baja California, Mexico

ISSN 2191-530X   ISSN 2191-5318 (electronic)
SpringerBriefs in Applied Sciences and Technology
ISSN 2625-3704   ISSN 2625-3712 (electronic)
SpringerBriefs in Computational Intelligence
ISBN 978-3-030-43949-1   ISBN 978-3-030-43950-7 (eBook)
https://doi.org/10.1007/978-3-030-43950-7

This Springer imprint is published by the registered company Springer Nature Switzerland AG
The registered company address is: Gewerbestrasse 11, 6330 Cham, Switzerland

Preface

This book focuses on the fields of fuzzy logic and metaheuristic algorithms, especially the harmony search algorithm and also considering the fuzzy control area. At the present time, there are several types of metaheuristics which have been used to solve various types of problems in the real world. These metaheuristics contain parameters that are usually fixed throughout the iterations. However, various techniques exist to dynamically adjust the parameters of an algorithm, such as probabilistic and fuzzy logic, among others.

The main contribution of this book is the proposed approach for solving this problem applied to the original harmony search algorithm using type-1, interval type-2 and generalized type-2 fuzzy logic. The proposed methodology was applied to the resolution of problems of classical benchmark mathematical functions, CEC 2015, CEC2017 functions and to the optimization of various fuzzy logic control cases. In this book we tested the proposed method using six benchmark problems; four are of the Mamdani type and two of the Sugeno type. The Mamdani problems are the following: the problem of filling a water tank, the problem of controlling the temperature of a shower, the problem of controlling the trajectory of an autonomous mobile robot and the problem of controlling the speed of an engine. The Sugeno problems are the following: the problem to control the balance of the bar and ball and the problem to control the balance of the inverted pendulum. When the interval and generalized type-2 fuzzy logic systems are implemented to model the behavior of the systems, the results show a better stabilization because the uncertainty analysis is better. For this reason, we consider in this book that the proposed method uses fuzzy systems, fuzzy controllers and the harmony search optimization algorithm to improve the behavior of complex control problems.

In Chap. 1, we begin by offering a brief introduction of the potential use of the optimization strategies in different real-world applications. We describe the use of the harmony search optimization algorithm using type-1, interval type-2 and generalized type-2 fuzzy logic systems for aggregation of results in problems of intelligent control of nonlinear plants. We also mention other possible applications of the proposed control approach.

We describe in Chap. 2 the basic concepts, notation and theory of the original harmony search algorithm. This chapter overviews the main definitions, equations and basic concepts, useful for the development of this research work.

We describe in Chap. 3 the proposed fuzzy harmony search algorithm (FHS) method that is based on the original harmony search algorithm (HS), and it shows in detail the equations used for the input and output of the fuzzy system that will be responsible for dynamically adjusting the parameters according to the iterations advanced.

Chapter 4 presents detailed case studies of mathematical functions and benchmark cases of control. It also includes the experimental results obtained with benchmark classic mathematical functions, CEC 2015, CEC 2017 and four cases of Mamdani control and two cases of Sugeno control, and also the statistical tests that validate the improvement of the proposed method are shown.

We describe in Chap. 5 the conclusions of this work, as well as some envisioned future research work. A new fuzzy harmony search algorithm optimization for benchmark mathematical functions and control was proposed, and then different cases of control were studied. Different control cases were presented. These details were obtained working first with their optimization using original harmony search algorithm and using type-1, interval type-2 and generalized type-2 fuzzy systems with the proposed method.

We end this preface of the book by giving thanks to all the people who have help or encourage us during the writing of this book. First of all, we would like to thank our professors and advisors for always supporting our work, and for motivating us to write our research work. We would also like to thank our colleagues working in Soft Computing, which are too many to mention each by their name. Of course, we need to thank our supporting agencies, CONACYT and TNM, in our country for their help during this project. We have to thank our institution, Tijuana Institute of Technology, for always supporting our projects. Finally, we thank our respective families for their continuous support during the time that we spend in this project.

Tijuana, Mexico   Prof. Dr. Fevrier Valdez
                                                                               Dr. Cinthia Peraza
                                                                       Prof. Dr. Oscar Castillo

Contents

1   **Introduction to Fuzzy Harmony Search** 1
    References 2

2   **Theory of the Original Harmony Search Method** 5
    2.1   Original Harmony Search (HS) 5
        2.1.1   Memory in the Harmony Search Algorithm 6
        2.1.2   Pitch Adjusting 6
        2.1.3   Randomization 6
    References 7

3   **Proposed Fuzzy Harmony Search Method** 9
    3.1   Mathematical Description 9
    Reference 11

4   **Study Cases to Test Fuzzy Harmony Search** 13
    4.1   Benchmark Mathematical Functions 13
        4.1.1   Classic Benchmark Mathematical Functions 13
        4.1.2   CEC 2015 Benchmark Mathematical Functions 24
        4.1.3   CEC 2017 Benchmark Mathematical Functions 32
    4.2   Benchmark Control Problems 40
        4.2.1   Mamdani Controller 40
        4.2.2   Sugeno Controller 53
    References 66

5   **Conclusions to Fuzzy Harmony Search** 69

Appendix 73

Index 83

Chapter 1
Introduction to Fuzzy Harmony Search

There are several metaheuristics currently being used to solve many real-life problems, such as problems of mathematical optimization [1–5], control problems [6–11], engineering problems [12–15] and scheduling problems [16–19], among many others. These metaheuristics usually have parameters that help the diversification and intensification properties in exploring of the search space, but use fixed parameters. To solve the problem of the fixed parameters within an algorithm, probabilistic methods [20–22] and fuzzy logic [23–28], among others, have been put forward. Fuzzy logic aims at creating mathematical approximations in the solution of certain types of problems because it provides the ability of modeling uncertainty, vagueness and imprecision present in the majority of real-world problems. It aims to produce accurate results from inaccurate data, which is why it is particularly useful in electronic or computational applications. It is defined as a mathematical system that models nonlinear functions, which converts inputs into outputs according to the logical approaches used by approximate reasoning. Fuzzy logic is very popular for its wide range of applications. In general, it is applied for both purposes, to control systems of complex industrial processes and to model any continuous system of engineering, physics, biology or economics. These concepts are applied to the decision-making process in a fuzzy logic controller, which has been elevated from a fuzzy type-1 system (T1-FLS), to become an interval type-2 fuzzy system (T2-FLS) where uncertainty could be incorporated directly into fuzzy sets [29–35]. When uncertainty is considered, system improvements are obtained, resilience against noise and handling of unknown data, and there are several tasks where this knowledge has been applied, as shown in [9, 36–42]. In generalized type-2 fuzzy set (GT2-FLS) the uncertainty is represented in volume and is able to handle greater uncertainty in the system. Some generalized type-2 advances in the literature are presented in [43–47].

F. Valdez et al., *General Type-2 Fuzzy Logic in Dynamic Parameter Adaptation for the Harmony Search Algorithm*, SpringerBriefs in Computational Intelligence, https://doi.org/10.1007/978-3-030-43950-7_1

The main contribution of this book is to solve the problem of using fixed parameters in the original harmony search algorithm [48, 49] which is based on jazz improvisation, by using a type-1 (T1-FHS), interval type-2 (IT2-FHS) and generalized type-2 (GT2-FHS) fuzzy logic systems that are responsible for dynamically adjusting the parameter throughout the iterations and are applied to the optimization of a classical and complex benchmark mathematical functions and various control cases. Experiments were performed without noise and with noise within the fuzzy logic controller for control cases.

The rest of the book is organized as follows: Chap. 2 introduces the original harmony search algorithm. In Chap. 3, the proposed method is discussed. The case study and optimization method are characterized in Chap. 4. Finally, Chap. 5 offers the conclusion.

References

1. Ochoa, P., Castillo, O., Soria, J.: Interval type-2 fuzzy logic dynamic mutation and crossover parameter adaptation in a fuzzy differential evolution method. In: Hadjiski, M., Atanassov, K.T. (eds.) Intuitionistic Fuzziness and Other Intelligent Theories and Their Applications, pp. 81–94. Springer International Publishing, Cham (2019)
2. Rodríguez, L., et al.: A fuzzy hierarchical operator in the grey wolf optimizer algorithm. Appl. Soft Comput. **57**, 315–328 (2017)
3. Bernal, E., Castillo, O., Soria, J., Valdez, F.: Imperialist competitive algorithm with dynamic parameter adaptation using fuzzy logic applied to the optimization of mathematical functions. Algorithms **10**(1), 18 (2017)
4. Ochoa, P., Castillo, O., Soria, J.: Interval type-2 fuzzy logic dynamic mutation and crossover parameter adaptation in a fuzzy differential evolution method. In: Hadjiski, M., Atanassov, K.T. (eds.) Intuitionistic Fuzziness and Other Intelligent Theories and Their Applications, vol. 757, pp. 81–94. Springer International Publishing, Cham (2019)
5. Barraza, J., Melin, P., Valdez, F., Gonzalez, C.: Fuzzy fireworks algorithm based on a sparks dispersion measure. Algorithms **10**(4), 83 (2017)
6. Amador-Angulo, L., Castillo, O.: A new fuzzy bee colony optimization with dynamic adaptation of parameters using interval type-2 fuzzy logic for tuning fuzzy controllers. Soft. Comput. **22**(2), 571–594 (2018)
7. Castillo, O., Valdez, F., Soria, J., Amador-Angulo, L., Ochoa, P., Peraza, C.: Comparative study in fuzzy controller optimization using bee colony, differential evolution, and harmony search algorithms. Algorithms **12**(1), 9 (2018)
8. Bernal, E., Castillo, O., Soria, J., Valdez, F.: Optimization of fuzzy controller using galactic swarm optimization with type-2 fuzzy dynamic parameter adjustment. Axioms **8**(1), 26 (2019)
9. Peraza, C., Valdez, F., Castro, J.R., Castillo, O.: Fuzzy dynamic parameter adaptation in the harmony search algorithm for the optimization of the ball and beam controller. Adv. Oper. Res. **2018**, 1–16 (2018)
10. Peraza, Cinthia, Valdez, Fevrier, Melin, Patricia: Optimization of intelligent controllers using a type-1 and interval type-2 fuzzy harmony search algorithm. Algorithms **10**(3), 82 (2017)
11. Caraveo, C., Valdez, F., Castillo, O.: Optimization of fuzzy controller design using a new bee colony algorithm with fuzzy dynamic parameter adaptation. Appl. Soft Comput. **43**, 131–142 (2016)

12. Dhiman, G., Kumar, V.: Spotted hyena optimizer for solving complex and non-linear constrained engineering problems. In: Yadav, N., Yadav, A., Bansal, J.C., Deep, K., Kim, J.H. (eds.) Harmony Search and Nature Inspired Optimization Algorithms, vol. 741, pp. 857–867. Springer Singapore, Singapore (2019)

13. El-Shorbagy, M.A., Farag, M.A., Mousa, A. A., El-Desoky, I.M.: A hybridization of sine cosine algorithm with steady state genetic algorithm for engineering design problems, pp. 143–155. Heidelberg: Springer Berlin Heidelberg, Berlin (2020)

14. Dhiman, G., Kumar, V.: Seagull optimization algorithm: theory and its applications for large-scale industrial engineering problems. Knowl.-Based Syst. **165**, 169–196, Feb (2019)

15. Lee, H.M., Jung, D., Sadollah, A., Lee, E.H., Kim, J.H.: Performance comparison of meta-heuristic optimization algorithms using water distribution system design benchmarks. In: Yadav, N., Yadav, A., Bansal, J.C., Deep, K., Kim, J.H. (eds.) Harmony Search and Nature Inspired Optimization Algorithms, vol. 741, pp. 97–104. Springer Singapore, Singapore (2019)

16. Halim, A.H., Ismail, I.: Combinatorial optimization: comparison of heuristic algorithms in travelling salesman problem. Arch. Comput. Methods Eng. **26**(2), 367–380 (2019)

17. Karagul, K., Sahin, Y., Aydemir, E., Oral, A.: A simulated annealing algorithm based solution method for a green vehicle routing problem with fuel consumption. In: Paksoy, T., Weber, G.-W., Huber, S. (eds.) Lean and Green Supply Chain Management, vol. 273, pp. 161–187. Springer International Publishing, Cham (2019)

18. Roy, B., Sen, A.K.: Meta-heuristic techniques to solve resource-constrained project scheduling problem. In: Bhattacharyya, S., Hassanien, A.E., Gupta, D., Khanna, A., Pan, I. (eds.) International Conference on Innovative Computing and Communications, vol. 56, pp. 93–99. Springer Singapore, Singapore (2019)

19. Pongchairerks, P.: A two-level metaheuristic algorithm for the job-shop scheduling problem. Complexity **2019**, 1–11 (2019)

20. Bansal, J.C., Singh, P.K., Pal, N.R. (eds.): Evolutionary and Swarm Intelligence Algorithms, vol. 779. Springer International Publishing, Cham (2019)

21. Dechter, R.: Reasoning with probabilistic and deterministic graphical models: exact algorithms, second edition. Synth. Lect. Artif. Intell. Mach. Learn. **13**(1), 1–199 (2019)

22. Wang, X., Chang, M.-C., Wang, L., Lyu, S.: Efficient algorithms for graph regularized plsa for probabilistic topic modeling. Pattern Recognit. **86**, 236–247 (2019)

23. Santiago, A., Dorronsoro, B., Nebro, A.J., Durillo, J.J., Castillo, O., Fraire, H.J.: A novel multi-objective evolutionary algorithm with fuzzy logic based adaptive selection of operators: fame. Inf. Sci. **471**, 233–251 (2019)

24. Olivas, F., Valdez, F., Melin, P., Sombra, A., Castillo, O.: Interval type-2 fuzzy logic for dynamic parameter adaptation in a modified gravitational search algorithm. Inf. Sci. **476**, 159–175 (2019)

25. Peraza, C., Valdez, F., Castillo, O.: Fuzzy harmony search algorithm using an interval type-2 fuzzy logic applied to benchmark mathematical functions. In: Hadjiski, M., Atanassov, K.T. (eds.) Intuitionistic Fuzziness and Other Intelligent Theories and Their Applications, vol. 757, pp. 13–28. Springer International Publishing, Cham (2019)

26. Ontiveros, E., Melin, P., Castillo, O.: Impact study of the footprint of uncertainty in control applications based on interval type-2 fuzzy logic controllers. In: Castillo, O., Melin, P., Kacprzyk, J. (eds.) Fuzzy Logic Augmentation of Neural and Optimization Algorithms: Theoretical Aspects and Real Applications, vol. 749, pp. 181–197. Springer International Publishing, Cham (2018)

27. Ontiveros, E., Melin, P., Castillo, O.: High order α-planes integration: a new approach to computational cost reduction of general type-2 fuzzy systems. Eng. Appl. Artif. Intell. **74**, 186–197 (2018)

28. Ontiveros-Robles, E., Melin, P., Castillo, O.: New methodology to approximate type-reduction based on a continuous root-finding karnik mendel algorithm. Algorithms **10**(3), 77 (2017)

29. Karnik, N.N., Mendel, J.M., Liang, Q.: Type-2 fuzzy logic systems. IEEE Trans. Fuzzy Syst. **7**(6), 643–658, Dec (1999)

30. Karnik, N.N., Mendel, J.M.: Centroid of a type-2 fuzzy set. Inf. Sci. **132**(1–4), 195–220 (2001)

31. Karnik, N.N., Mendel, J.M.: Operations on type-2 fuzzy sets. Fuzzy Sets Syst. **122**(2), 327–348 (2001)
32. Mendel, J.M.: Advances in type-2 fuzzy sets and systems. Inf. Sci. **177**(1), 84–110 (2007)
33. Ruiz-Garcia, G., Hagras, H., Pomares, H., Rojas, I.: Towards a fuzzy logic system based on general forms of interval type-2 fuzzy sets. IEEE Trans. Fuzzy Syst., 1–1 (2019)
34. Wu, D., Mendel, J.M.: Recommendations on designing practical interval type-2 fuzzy systems. Eng. Appl. Artif. Intell. **85**, 182–193 (2019)
35. Mendel, J.M.: Computing derivatives in interval type-2 fuzzy logic systems. IEEE Trans. Fuzzy Syst. **12**(1), 84–98 (2004)
36. Olivas, F., Valdez, F., Castillo, O.: Comparison of bio-inspired methods with parameter adaptation through interval type-2 fuzzy logic. In: Castillo, O., Melin, P., Kacprzyk, J. (eds.) Fuzzy Logic Augmentation of Neural and Optimization Algorithms: Theoretical Aspects and Real Applications, vol. 749, pp. 39–53. Springer International Publishing, Cham (2018)
37. Ramirez, E., Melin, P., Prado-Arechiga, G.: Hybrid model based on neural networks, type-1 and type-2 fuzzy systems for 2-lead cardiac arrhythmia classification. Expert Syst. Appl. **126**, 295–307 (2019)
38. Castillo, O., Atanassov, K.: Comments on fuzzy sets, interval type-2 fuzzy sets, general type-2 fuzzy sets and intuitionistic fuzzy sets. In: Melliani, S., Castillo, O. (eds.) Recent Advances in Intuitionistic Fuzzy Logic Systems, vol. 372, pp. 35–43. Springer International Publishing, Cham (2019)
39. Guzmán, J., Miramontes, I., Melin, P., Prado-Arechiga, G.: Optimal genetic design of type-1 and interval type-2 fuzzy systems for blood pressure level classification. Axioms **8**(1), 8 (2019)
40. Miramontes, I., Guzman, J., Melin, P., Prado-Arechiga, G.: Optimal design of interval type-2 fuzzy heart rate level classification systems using the bird swarm algorithm. Algorithms **11**(12), 206 (2018)
41. Castro, J.R., Castillo, O., Melin, P.: An interval type-2 fuzzy logic toolbox for control applications. In: 2007 IEEE International Fuzzy Systems Conference, pp. 1–6. London, UK (2007)
42. Sanchez, M.A., Castillo, O., Castro, J.R.: Generalized type-2 fuzzy systems for controlling a mobile robot and a performance comparison with interval type-2 and type-1 fuzzy systems. Expert Syst. Appl. **42**(14), 5904–5914 (2015)
43. Gonzalez, C.I., Melin, P., Castro, J.R., Castillo, O.: Edge detection methods based on generalized type-2 fuzzy logic systems. In: Edge Detection Methods Based on Generalized Type-2 Fuzzy Logic, pp. 21–35. Springer International Publishing, Cham (2017)
44. Castillo, O., Amador-Angulo, L.: A generalized type-2 fuzzy logic approach for dynamic parameter adaptation in bee colony optimization applied to fuzzy controller design. Inf. Sci., Oct (2017)
45. Gonzalez, C.I., Melin, P., Castro, J.R., Castillo, O.: Generalized type-2 fuzzy edge detection applied on a face recognition system. In: Edge Detection Methods Based on Generalized Type-2 Fuzzy Logic, pp. 37–41. Springer International Publishing, Cham (2017)
46. Hao, M., Mendel, J.M.: Similarity measures for general type-2 fuzzy sets based on the α-plane representation. Inf. Sci. **277**, 197–215 (2014)
47. Liu, F.: An efficient centroid type-reduction strategy for general type-2 fuzzy logic system. Inf. Sci. **178**(9), 2224–2236 (2008)
48. Geem, Z.W., Kim, J.H., Loganathan, G.V.: A new heuristic optimization algorithm: harmony search. Simulation **76**(2), 60–68, Feb (2001)
49. Lee, K.S., Geem, Z.W.: A new meta-heuristic algorithm for continuous engineering optimization: harmony search theory and practice. Comput. Methods Appl. Mech. Eng. **194**(36–38), 3902–3933 (2005)

Chapter 2
Theory of the Original Harmony Search Method

In this chapter we present some basic concepts about the search algorithm of the original harmony to better understand the idea and context of this book.

2.1 Original Harmony Search (HS)

The harmony search algorithm (HS) is a recently proposed metaheuristic search technique that mimics the process of music improvisation. The concept of HS was introduced by Zong Woo Geem in 2001, and it is inspired by the improvisation of music, specifically referred to jazz. There are different variants of this algorithm as they have been created to improve it or for solving specific problems [1–3]. The HS algorithm and its variants have been used in multiple areas of applications, such as engineering problems [4–7], public key cryptography [8], route problem [9] and medical area [10], among others. This algorithm can be explained in more detail with the process of improvisation that a musician uses, which consists of three options:

1. Play any famous piece of music.
2. Play something similar to a known piece.
3. Compose a new or random.

If we formalize these three options for optimization, we have three corresponding components: using the harmony memory rate (*HMR*), pitch adjusting rate (*PArate*) and randomization.

The harmony search algorithm has three operators. The first, harmony memory accepting rate (*HMR*) represents harmony exploitation of the search space and the literature tells us that the range is from 0 to 1, but for best results we can use the range between 0.7 and 0.95.

2.1.1 Memory in the Harmony Search Algorithm

The usage of harmony memory is important as it is similar to choosing the best fit individuals in genetic algorithms. This will ensure that the best harmonies will be carried over to the new harmony memory. In order to use this memory more effectively, we can assign a parameter $r_{accept} \in [0, 1]$, called harmony memory accepting rate (*HMR*). If this rate is too low, only a few of the best harmonies are selected and HS may converge too slowly.

$$r_{accept} \in [0, 1] \tag{2.1}$$

If this rate is extremely high (near 1), almost all the harmonies are used in the harmony memory, then other harmonies are not explored well, leading to potentially wrong solutions. Therefore, typically, r_{accept} is in the interval [0.7–0.95].

The two other operators are the pitch adjusting (*PArate*) settings and randomization. These two serve to control the level of exploration in the search space and the range used is from 0 to 1. However, the literature indicates that we can use a smaller range of 0.1–0.5 for best results, and in this case, we used the range from 0.7 to 1.

2.1.2 Pitch Adjusting

To adjust the pitch slightly in the second component, we have to use a method that can adjust the frequency efficiently. In theory, the pitch can be adjusted linearly or nonlinearly, but in practice the linear adjustment is used in most of the cases. If x_{old} is the current solution (or pitch), then the new solution (pitch) x_{new} is generated by

$$x_{new} = x_{old} + b_p(2\,rand - 1) \tag{2.2}$$

where "*rand*" is a random number drawn from a uniform distribution [0, 1]. Here b_p is the bandwidth, which controls the local range of pitch adjustment. In fact, we can note that the pitch adjusting (2.2) is a random walk.

Pitch adjustment is similar to the mutation operator in genetic algorithms. We can assign a pitch adjusting rate (*PArate*) to control its degree of adjustment. If *PArate* is too low, then there is rarely any change. If it is too high, then the algorithm may not converge at all. Thus, we usually use $r_{pa} = 0.1 \sim 0.5$.

2.1.3 Randomization

The third component is the randomization, which is basically used to increase the diversity of the solutions. Adjusting the pitch has a similar role, but it is limited to certain local pitch adjustment and thus corresponds to a local search. The use of randomization can drive the system further to explore various regions with high solutions diversity so as to find the global optimal. So we have:

$$P_a = P_{lowerlimit} + P_{range} * rand \tag{2.3}$$

where $P_{range} = P_{upperlimit} - P_{lowerlimit}$. Here "*rand*" is a random number generator in the range between 0 and 1.

$$P_{range} = P_{upperlimit} - P_{lowerlimit} \tag{2.4}$$

The three components in harmony search can be summarized in:

$$P_{random} = 1 - r_{accept}$$

And the actual probability of pitch adjusting is

$$P_{pitch} = r_{accept} * r_{pa} \tag{2.5}$$

References

1. Mahdavi, M., Fesanghary, M., Damangir, E.: An improved harmony search algorithm for solving optimization problems. Appl. Math. Comput. **188**(2), 1567–1579 (2007)
2. Wang, C.-M., Huang, Y.-F.: Self-adaptive harmony search algorithm for optimization. Expert Syst. Appl. **37**(4), 2826–2837 (2010)
3. Geem, Z.W.: Novel derivative of harmony search algorithm for discrete design variables. Appl. Math. Comput. **199**(1), 223–230 (2008)
4. Yi, J., Li, X., Chu, C.-H., Gao, L.: Parallel chaotic local search enhanced harmony search algorithm for engineering design optimization. J. Intell. Manuf. **30**(1), 405–428 (2019)
5. Yi, J., Gao, L., Li, X., Shoemaker, C.A., Lu, C.: An on-line variable-fidelity surrogate-assisted harmony search algorithm with multi-level screening strategy for expensive engineering design optimization. Knowl.-Based Syst. **170**, 1–19, Apr (2019)
6. Schaedler de Almeida, F.: Optimization of laminated composite structures using harmony search algorithm. Compos. Struct. **221**, 110852 (2019)
7. Nazari-Heris, M., Mohammadi-Ivatloo, B., Asadi, S., Geem, Z.W.: Large-scale combined heat and power economic dispatch using a novel multi-player harmony search method. Appl. Therm. Eng. **154**, 493–504 (2019)
8. Mitra, S., Mahapatra, G., Balas, V.E., Chattaraj, R.: Public key cryptography using harmony search algorithm. In: Deb, D., Balas, V.E., Dey, R. (eds.) Innovations in Infrastructure, vol. 757, pp. 1–11. Springer Singapore, Singapore (2019)
9. Boryczka, U., Szwarc, K.: The Harmony Search algorithm with additional improvement of harmony memory for Asymmetric Traveling Salesman Problem. Expert Syst. Appl. **122**, 43–53 (2019)
10. Selvakumar, S., Abdullah, A.S., Suganya, R.: Decision support system for type II diabetes and its risk factor prediction using bee-based harmony search and decision tree algorithm. Int. J. Biomed. Eng. Technol. **29**(1), 46 (2019)

Chapter 3
Proposed Fuzzy Harmony Search Method

The proposed method is based on the HS algorithm, and the *HMR* and *PArate* parameters introduced in the improvisation process help the algorithm to find globally and locally improved solutions. The traditional HS algorithm uses fixed values for *HMR* and *PArate* parameters whereas the improvement suggested in [1] varies the PArate and b_w parameters throughout the iterations. Inspired by their modification, we propose here a new fuzzy harmony search algorithm (FHS) by using a type-1 (T1-FHS), interval type-2 (IT2-FHS) and generalized type-2 (GT2-FHS) fuzzy logic systems.

3.1 Mathematical Description

This chapter describes the formal mathematical definition of the proposed FHS method. The mathematical description for FHS is defined as follows:

Equation 2.1 shows the fundamental part of the traditional HS algorithm, where the harmony memory accepting (*HMR*) is represented by a constant value. Equation 2.2 shows the fundamental part of the traditional HS algorithm, where the pitch adjusting (*PArate*) is represented by a constant value. Equations 3.1 and 3.2 describe the different ways the basic equations of HS are modified for achieving the goal, and then, convert part of it in fuzzy parameters. We can notice the differences between both equations, in that *HMR* in Eq. 3.1 and *PArate* in Eq. 3.2 are those that change, because an essential part of the proposed method lies in these two parameters. Traditionally, these two parameters are constant; in this case, due to the importance of these two parameters, we decided to make these two as fuzzy parameters. Therefore, for designing the fuzzy systems, which dynamically adjust the *HMR* and *PArate* parameters, the measure of percentage of iterations is considered as input. All the fuzzy systems are of Mamdani type because it is more common to use them in this type of fuzzy control and the defuzzification method is the centroid. In this case, we are using this type of defuzzification because in other papers we have achieved good results with this method.

In this case, these values are considered fuzzy because they are changing dynamically when the FHS is running, and are defined by Eqs. 3.1 and 3.2, where HMR and PArate are changing values in the range [0,1].

$$HMR = \frac{\sum_{i=1}^{r_{hmr}} \mu_i^{hmr} (hmr_{1i})}{\sum_{i=1}^{r_{hmr}} \mu_i^{hmr}} \tag{3.1}$$

where HMR is the harmony memory accepting rate; r_{hmr} is the number of rules of the fuzzy system corresponding to hmr; hmr_{1i} is the output result for rule i corresponding to hmr; and μ_i^{hmr} is the membership function of rule i corresponding to hmr.

$$PArate = \frac{\sum_{i=1}^{r_{PArate}} \mu_i^{PArate} (PArate_{1i})}{\sum_{i=1}^{r_{PArate}} \mu_i^{PArate}} \tag{3.2}$$

where PArate is the pitch adjusting considering rate; r_{PArate} is the number of rules of the fuzzy system corresponding to PArate; $PArate_{1i}$ is the output result for rule i corresponding to PArate; and μ_i^{PArate} is the membership function of rule i corresponding to PArate.

In addition, it is also found that an algorithm performance measure, such as the iterations, needs to be considered in the parameter adaptation. In this chapter all the above are taken into consideration for the fuzzy systems to modify the HMR and PArate parameters by dynamically changing these parameters in each iteration of the algorithm.

In this phase we consider a percentage of iterations elapsed to determine the values of HMR and PArate. If a low percentage of iterations elapsed, the HMR and PArate parameters would take a low value in the range, and this occurs in order to enable a wide search in the search space, as a kind of exploration. On the contrary, if a high percentage of iterations has occurred, it would give the HMR and PArate parameters a high value in the indicated range, and this occurs in order to draw a kind of more intense exploitation within the search space in iterations advanced. Now, if the percentage of elapsed iterations is medium, then a value of HMR and PArate as a medium would occur, with the aim of expanding the above criteria. To represent this idea, we use the following equation:

$$Iteration = \frac{Current\ Iteration}{Maximun\ of\ iterations} \tag{3.3}$$

The diagram of the proposed method is illustrated in Fig. 3.1.

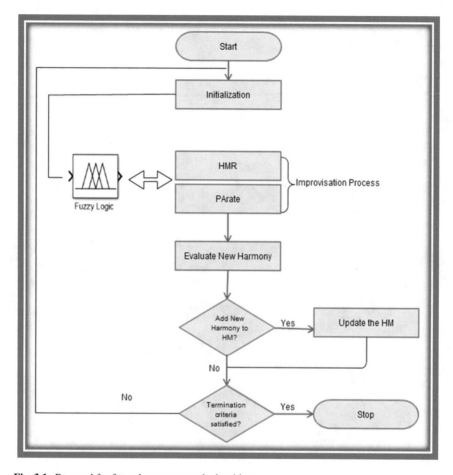

Fig. 3.1 Proposal for fuzzy harmony search algorithm

Reference

1. Mahdavi, M., Fesanghary, M., Damangir, E.: An improved harmony search algorithm for solving optimization problems. Appl. Math. Comput. **188**(2), 1567–1579 (2007)

Chapter 4
Study Cases to Test Fuzzy Harmony Search

In this chapter we present the different case studies used and analyzed to prove the performance and efficiency of the metaheuristics developed in this book. In Chap. 3, the general methodology of the proposal is shown, and this methodology was applied to the following cases of study: benchmark mathematical functions and control problems.

4.1 Benchmark Mathematical Functions

There are several mathematical functions to validate the performance of the algorithms, and these are divided into unimodal, multimodal and compound hybrid, among others. They are classified due to their degree of complexity, and in the following sections the functions that are studied in each case are shown.

4.1.1 Classic Benchmark Mathematical Functions

In the previous work [1] it was concluded that adjusting the HMR parameter along the iterations in the harmony search algorithm achieved good results applied to classic mathematical benchmark functions with 10 dimensions.

The classic mathematical benchmark functions used are shown in Table 4.1. The function, the search space, the global minimum and the equation for each function are indicated.

© The Author(s), under exclusive license to Springer Nature Switzerland AG 2020
F. Valdez et al., *General Type-2 Fuzzy Logic in Dynamic Parameter Adaptation for the Harmony Search Algorithm*, SpringerBriefs in Computational Intelligence,
https://doi.org/10.1007/978-3-030-43950-7_4

Table 4.1 Classic benchmark mathematical functions

Function	Search domain	f min	Equation
Sphere	$[-5.12, 5.12]^n$	0	$f(x) = \sum\limits_{i=1}^{n} x_i^2$
Rosenbrock	$[-5, 10]^n$	0	$f(x) =$ $\sum\limits_{i=1}^{n-1}\left[100\left(x_{i-1} - x_i^2\right)^2 + (1 - x_1)^2\right]$
Rastrigin	$[-5.12, 5.12]^n$	0	$f(x) =$ $10n + \sum\limits_{i=1}^{n}\left[x_i^2 - 10\cos(2x_i)\right]$
Ackley	$[-32.768, 32.768]^n$	0	$f(x) = -a \cdot \exp\left(-b \cdot \sqrt{\frac{1}{n}\sum\limits_{i-1}^{n} x_i^2}\right) -$ $\exp(\frac{1}{n}\sum\limits_{i-1}^{n}\cos(cx_i)) + a + \exp(a)$
Zakharov	$[-5, 10]^n$	0	$f(x) =$ $\sum\limits_{i=1}^{n} x_i^2 + (\sum\limits_{i+1}^{n} 0.5ix_i)^2 + (\sum\limits_{i=1}^{n} 0.5ix_i)^4$
Sum squares	$[-10, 10]^n$	0	$f(x) = \sum\limits_{i=1}^{n} ix_i^2$
Griewank	$[-600, 600]^n$	0	$f(x) = \sum\limits_{i=1}^{n} \frac{x_i^2}{4000} - \prod\limits_{i=1}^{n} \cos\left(\frac{x_i}{\sqrt{i}}\right) + 1$
Schwefel	$[-500, 500]^n$	0	$f(x) =$ $418.9829d - \sum\limits_{i=1}^{n} x_i \sin(\sqrt{|x_i|})$
Six-hump camel back	$[-3, 3]^n$	−1.0316	$f(x) = \left(4 - 2.1x_1^2 + \frac{x_1^4}{3}\right)x_1^2 +$ $x_1x_2 + \left(-4 + 4x_2^2\right)x_2^2$
Rotated hyper-ellipsoid	$[-65.536, 65.536]^n$	0	$f(x) = \sum\limits_{i=1}^{d} \sum\limits_{j=1}^{i} x_j^2$

4.1.1.1   Design of the Type-1 (T1-FHS), Interval Type-2 and Generalized Type-2 Fuzzy Harmony Search System (IT2-FHS)

The proposed fuzzy system (T1-FHS) is composed of using terms of type-1 fuzzy sets [1] granulated into the three triangular membership functions (MFs) in the input and the output, and these triangular membership functions are represented by Eq. 4.1.

Table 4.2 Type-1 membership functions for the input and output variables

Low	$\mu Low(x) = \begin{cases} 0, & x \leq -0.5 \\ \frac{x+0.5}{0+0.5}, & -0.5 \leq x \leq 0 \\ \frac{0.5-x}{0.5-0}, & 0 \leq x \leq 0.5 \\ 0, & 0.5 \leq x \end{cases}$
Medium	$\mu Medium(x) = \begin{cases} 0, & x \leq 0 \\ \frac{x-0}{0.5-0}, & 0 \leq x \leq 0.5 \\ \frac{1-x}{1-0.5}, & 0.5 \leq x \leq 1 \\ 0, & 1 \leq x \end{cases}$
High	$\mu High(x) = \begin{cases} 0, & x \leq 0.5 \\ \frac{x-0.5}{1-0.5}, & 0.5 \leq x \leq 1 \\ \frac{1.5-x}{1.5-1}, & 1 \leq x \leq 1.5 \\ 0, & 1.5 \leq x \end{cases}$

$$f(x; a, b, c) = \begin{cases} 0, & x \leq a \\ \frac{x-a}{b-a}, & a \leq x \leq b \\ \frac{c-x}{c-b}, & b \leq x \leq c \\ 0, & c \leq x \end{cases} \tag{4.1}$$

To build a type-1 fuzzy system, the input and output linguistic variables are determined; in this case, there is one input and one output. The input variable is the *Iteration*, and this variable has the following membership functions (MFs): low, medium and high. The fuzzy system has a single output called HMR, which refers to the harmony memory accepting parameter, and this variable has the following MFs: low, medium and high. These MFs are listed in Table 4.2.

This T1-FHS is illustrated in Fig. 4.1.

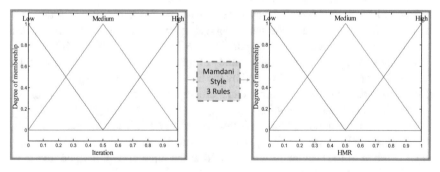

Fig. 4.1 Graphical representation of the type-1 proposed method (FHS)

```
1.  If (Iteration is Low) then (HMR is Low)
2.  If (Iteration is Medium) then (HMR is Medium)
3.  If (Iteration is High) then (HMR is High)
```

Fig. 4.2 Rules for fuzzy system for increasing (T1-FHS)

To design the rules of this fuzzy system, it was decided that in early iterations the HS algorithm must explore and eventually exploit.

The rules for the fuzzy system for increasing the HMR parameter are shown in Fig. 4.2.

The proposed interval type-2 fuzzy system (IT2-FHS) is composed of using terms of interval type-2 fuzzy sets, granulated into the three triangular MFs with uncertain a in the input and the output, and these triangular MFs are represented by Eq. 4.2:

$$\tilde{\mu}(x) = [\underline{\mu}(x), \bar{\mu}(x)] = \text{itritype } 2(x, [a_1, b_1, c_1, a_2, b_2, c_2]) \tag{4.2}$$

where $a_1 < a_2, b_1 < b_2, c_1 < c_2$

$$\mu_1(x) = \max\left(\min\left(\frac{x - a_1}{b_1 - a_1}, \frac{c_1 - x}{c_1 - b_1}\right), 0\right)$$

$$\mu_2(x) = \max\left(\min\left(\frac{x - a_2}{b_2 - a_2}, \frac{c_2 - x}{c_2 - b_2}\right), 0\right)$$

$$\bar{\mu}(x) = max(\mu_1(x), \mu_2(x)) \forall x \notin (b1, b2)$$

$$\bar{\mu}(x) = 1 \forall x \in (b1, b2)$$

$$\underline{\mu}c(x) = min(\mu_1(x), \mu_2(x))$$

where the a_1, b_1 and c_2 parameters are for the upper MFs and a_2, b_2 and c_1 are for the lower MFs. The following equations represent the knowledge of the input and output variables of the interval type-2 fuzzy system.

To build an interval type-2 fuzzy system, we used the same general structure of the type-1 fuzzy system and the same linguistic variables mentioned above; in this case, we have one input and one output. These MFs are listed in Table 4.3.

The rules are designed based on the study of parameters of the algorithm, so that in the initial iterations it will explore and by the final iterations it will exploit the search space, and in this case the rules are on an increase fashion, as shown in Fig. 4.2. This IT2-FHS is illustrated in Fig. 4.3.

The proposed generalized type-2 fuzzy system (GT2-FHS) is composed of using terms of generalized type-2 fuzzy sets, granulated into the three triangular MFs with Gaussian in the secondary membership in the input and the output, and these triangular MFs are represented by Eq. 4.3:

Table 4.3 Interval type-2 membership functions for the input and output variables

Low	
	$\mu_1(x) = \max\left(\min\left(\dfrac{x - 0.58}{-0.08 - 0.58}, \dfrac{0.41 - x}{0.41 + 0.08}\right), 0\right)$
	$\mu_2(x) = \max\left(\min\left(\dfrac{x + 0.41}{0.08 + 0.41}, \dfrac{0.58 - x}{0.58 - 0.08}\right), 0\right)$
	$\bar{\mu}(x) = \max(\mu_1(x), \mu_2(x)) \forall x \notin (-0.08, 0.08)$
	$\bar{\mu}(x) = 1 \forall x \in (-0.08, 0.08)$
	$\underline{\mu}(x) = \min(\mu_1(x), \mu_2(x))$
Medium	
	$\mu_1(x) = \max\left(\min\left(\dfrac{x + 0.83}{0.41 + 0.83}, \dfrac{0.91 - x}{0.91 - 0.41}\right), 0\right)$
	$\mu_2(x) = \max\left(\min\left(\dfrac{x - 0.08}{0.58 - 0.08}, \dfrac{1.08 - x}{1.08 - 0.58}\right), 0\right)$
	$\bar{\mu}(x) = \max(\mu_1(x), \mu_2(x)) \forall x \notin (0.41, 0.58)$
	$\bar{\mu}(x) = 1 \forall x \in (0.41, 0.58)$
	$\underline{\mu}(x) = min(\mu_1(x), \mu_2(x))$
High	
	$\mu_1(x) = \max\left(\min\left(\dfrac{x - 0.41}{0.91 - 0.41}, \dfrac{1.41 - x}{1.41 - 0.91}\right), 0\right)$
	$\mu_2(x) = \max\left(\min\left(\dfrac{x - 0.58}{1.08 - 0.58}, \dfrac{1.58 - x}{1.58 - 1.08}\right), 0\right)$
	$\bar{\mu}(x) = max(\mu_1(x), \mu_2(x)) \forall x \notin (0.91, 1.08)$
	$\bar{\mu}(x) = 1 \forall x \in (0.91, 1.08)$
	$\underline{\mu}(x) = min(\mu_1(x), \mu_2(x))$

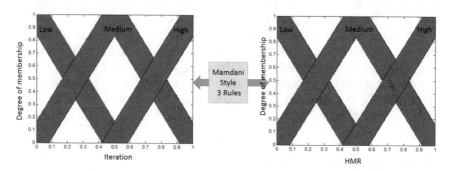

Fig. 4.3 Graphical representation of the interval type-2 proposed method (IT2-FHS)

$$\mu(x, u) = trigausstype2(x, u, [a_1, b_1, c_1, a_2, b_2, c_2, \rho])$$

$$\mu(x, u) = exp\left[-\frac{1}{2}\left(\frac{u - p_x}{\sigma_u}\right)^2\right] \text{ where}$$

$$\mu_1(x) = max\left(min\left(\frac{x - a_1}{b_1 - a_1}, \frac{c_1 - x}{c_1 - b_1}\right), 0\right) \text{ and}$$

$$\mu_2(x) = max\left(min\left(\frac{x - a_2}{b_2 - a_2}, \frac{c_2 - x}{c_2 - b_2}\right), 0\right),$$

$$\bar{\mu}(x) = \begin{cases} max(\mu_1(x), \mu_2(x)) \ \forall x \notin (b1, b2) \\ \qquad\qquad 1 \qquad\qquad \forall x \in (b1, b2) \end{cases}$$

$$\underline{\mu}(x) = min(\mu_1(x), \mu_2(x))$$

$$p_x = max\left(min\left(\frac{x - a_x}{b_x - a_x}, \frac{c_x - x}{c_x - b_x}\right), 0\right) \qquad\qquad (4.3)$$

where $a_x = \frac{a_1+a_2}{2}, b_x = \frac{b_1+b_2}{2}, c_x = \frac{c_1+c_2}{2}$

$$\delta = \bar{\mu}(x) - \underline{\mu}(x)$$

$$\sigma_u = \frac{1+\rho}{2\sqrt{3}}\delta + \varepsilon$$

where the a_1, b_1 and c_1 parameters are for the upper MFs and a_2, b_2 and c_2 are for the lower MFs, respectively; where ρ is fraction of uncertainty of the support for the secondary membership function. The following equations represent the knowledge of the input and output of the generalized type-2 fuzzy system.

To build a generalized type-2 fuzzy system, we used the same general structure of the type-1 fuzzy system and the same linguistic variables mentioned above; in this case, we have one input and one output. These MFs are listed in Table 4.4.

Table 4.4 shows the equations for constructing the MFs of the input and output of the generalized type-2 fuzzy system. The graphical representation of the fuzzy system is determined by Fig. 4.4.

4.1.1.2 Experiments and Results

For the experiments we use the mathematical functions presented in Table 4.1 and the parameters used for the experiment are shown in Table 4.5.

A total of 30 experiments were performed with 10 dimensions for each mathematical function and for each method: IT2-FHS with dynamic parameter adaptation of *HMR* parameter and IT2-FHS with dynamic parameter adaptation of *PArate* parameter. The results obtained are shown in Table 4.6.

Table 4.6 shows the results obtained when using the *HMR* and *PArate* parameter setting with T1-FHS and IT2-FHS for each mathematical function, and it can be seen

Table 4.4 Generalized type-2 membership functions for the input and output variables

Low	
	$\mu_1(x) = \max\left(\min\left(\dfrac{x+0.5807}{-0.0806+0.5807}, \dfrac{0.4193-x}{0.4193+0.0806}\right), 0\right)$ and
	$\mu_2(x) = \max\left(\min\left(\dfrac{x+0.4141}{0.0859+0.4141}, \dfrac{0.5859-x}{0.5859-0.0859}\right), 0\right)$
	$\bar{\mu}(x) = \begin{cases} max(\mu_1(x), \mu_2(x)) \ \forall x \notin (-0.0868, 0.0859) \\ 1 \qquad\qquad \forall x \in (-0.0868, 0.0859) \end{cases}$
	$\underline{\mu}(x) = min(\mu_1(x), \mu_2(x))$
	$p_x = \max\left(\min\left(\frac{x-a_x}{b_x-a_x}, \frac{c_x-x}{c_x-b_x}\right), 0\right),$ where
	$a_x = \frac{-0.5807-0.4141}{2}, b_x = \frac{-0.0806-0.0859}{2}, c_x = \frac{0.4193+0.5859}{2}$
	$\delta = \bar{\mu}(x) - \underline{\mu}(x)$
	$\sigma_u = \dfrac{1+\rho}{2\sqrt{3}}\delta + \varepsilon$
	where $\rho = 0.5$
Medium	
	$\mu_1(x) = \max\left(\min\left(\dfrac{x+0.0833}{0.4167+0.0833}, \dfrac{0.9167-x}{0.9167-0.4167}\right), 0\right)$ and
	$\mu_2(x) = \max\left(\min\left(\dfrac{x-0.0833}{0.5833-0.0833}, \dfrac{1.083-x}{1.083-0.5833}\right), 0\right)$
	$\bar{\mu}(x) = \begin{cases} max(\mu_1(x), \mu_2(x)) \ \forall x \notin (0.4167, 0.5833) \\ 1 \qquad\qquad \forall x \in (0.4167, 0.5833) \end{cases}$
	$\underline{\mu}(x) = min(\mu_1(x), \mu_2(x))$
	$p_x = \max\left(\min\left(\frac{x-a_x}{b_x-a_x}, \frac{c_x-x}{c_x-b_x}\right), 0\right),$ where
	$a_x = \frac{-0.0833+0.0833}{2}, b_x = \frac{0.4167+0.5833}{2}, c_x = \frac{0.9167+1.083}{2}$
	$\delta = \bar{\mu}(x) - \underline{\mu}(x)$
	$\sigma_u = \dfrac{1+\rho}{2\sqrt{3}}\delta + \varepsilon$
	where $\rho = 0.5$
High	
	$\mu_1(x) = \max\left(\min\left(\dfrac{x-0.422}{0.922-0.422}, \dfrac{1.422-x}{1.422-0.922}\right), 0\right)$ and
	$\mu_2(x) = \max\left(\min\left(\dfrac{x-0.5886}{1.088-0.5886}, \dfrac{1.588-x}{1.588-1.088}\right), 0\right)$
	$\bar{\mu}(x) = \begin{cases} max(\mu_1(x), \mu_2(x)) \ \forall x \notin (0.922, 1.088) \\ 1 \qquad\qquad \forall x \in (0.922, 1.088) \end{cases}$
	$\underline{\mu}(x) = min(\mu_1(x), \mu_2(x))$
	$p_x = \max\left(\min\left(\frac{x-a_x}{b_x-a_x}, \frac{c_x-x}{c_x-b_x}\right), 0\right),$ where
	$a_x = \frac{0.422+0.5886}{2}, b_x = \frac{0.922+1.088}{2}, c_x = \frac{1.422+1.588}{2}$
	$\delta = \bar{\mu}(x) - \underline{\mu}(x)$
	$\sigma_u = \dfrac{1+\rho}{2\sqrt{3}}\delta + \varepsilon$
	where $\rho = 0.5$

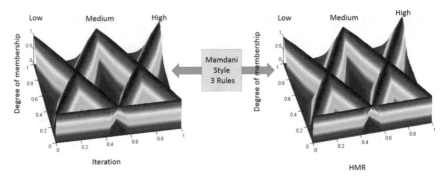

Fig. 4.4 Graphical representation of the generalized type-2 proposed method (GT2FHS)

Table 4.5 Parameters considered for each method

Parameter	Simple HS	T1-FHS	IT2-FHS	GT2-FHS
Harmonies	4–40	4–40	4–40	4–40
Iterations	1000	1000	1000	1000
HMR	0.95	Dynamic	Dynamic	Dynamic
PArate	0.7	0.7	0.7	0.7

Table 4.6 Results obtained for each method

Function	HS	T1-FHS [2] HMR parameter	IT2-FHS HMR [2] parameter	T1-FHS PArate parameter [3]	IT2-FHS PArate parameter [3]
Sphere	6.57E–05	6.33E–07	1.44E–10	1.38E–05	1.27E–10
Rosenbrock	4.88E–04	1.36E–08	2.52E–12	1.76E–11	3.42E–04
Ackley	1.18E–04	7.87E–05	9.28E–09	9.54E–06	6.50E–09
Rastrigin	2.26E–08	5.28E–05	9.55E–06	1.07E–08	1.05E–08
Zakharov	2.71E–04	1.25E–04	1.41E–05	4.72E–05	2.30E–03
Sum squared	3.59E–05	3.72E–06	2.09E–06	2.52E–06	7.43E–06

that when the HMR parameter is adjusted dynamically with interval type-2 fuzzy logic better results are obtained in all the mathematical functions.

The second comparison is against Wang et al. [4], where they proposed a new way to the harmony memory to automatically adjust parameter values.

In addition, the pseudo-random number generator is also replaced by the low-discrepancy sequences for initialization of the harmony memory. We use the same constrains used in [4], and in Tables 4.7 and 4.8 the results from [4] are presented.

From Table 4.7, the mean and standard deviation of the benchmark function optimization results for 30 dimensions for each mathematical function can be appreciated, and we use the same constrains used in [4]. The results were averaged over

Table 4.7 Comparison against Wang et al. Results for 30 dimensions [4]

30 Dimensions						
HS50 [4]			IT2-FHS HMR parameter		GT2-FHS	
Sphere	Mean	6.92E–07	Mean	3.52E–10	Mean	1.71E–10
	S.D.	1.10E–06	S.D.	3.85E–10	S.D.	2.99E–10
Rosenbrock	Mean	2.65E+01	Mean	3.12E–10	Mean	2.06E–10
	S.D.	5.68E–01	S.D.	3.42E–10	S.D.	3.12E–10
Ackley	Mean	7.81E–04	Mean	8.47E–07	Mean	3.66E–04
	S.D.	4.66E–04	S.D.	1.14E–06	S.D.	5.17E–04
Griewank	Mean	8.45E–05	Mean	4.58E–03	Mean	9.53E–02
	S.D.	2.38E–04	S.D.	2.11E–02	S.D.	6.80E–02

Table 4.8 Comparison against Wang et al. Results for 100 dimensions [4]

100 Dimensions						
HS50 [4]			IT2-FHS HMR parameter		GT2-FHS	
Sphere	Mean	1.53E–02	Mean	3.79E–10	Mean	2.39E–11
	S.D.	1.18E–02	S.D.	4.74E–10	S.D.	4.52E–11
Rosenbrock	Mean	9.60E+01	Mean	4.07E–10	Mean	1.75E–05
	S.D.	5.40E–01	S.D.	4.01E–10	S.D.	2.73E–05
Ackley	Mean	2.92E–02	Mean	7.46E–07	Mean	4.00E–04
	S.D.	9.81E–03	S.D.	8.50E–07	S.D.	6.79E–04
Griewank	Mean	6.66E–03	Mean	1.40E–02	Mean	5.81E–02
	S.D.	3.81E–03	S.D.	3.63E–02	S.D.	7.21E–02

30 runs. We can see that you between more complexes is the problem the type-2 algorithm sometimes get better results. Table 4.8 shows the simulation results for the 100 dimensions in each function.

From Table 4.8 the mean and standard deviation of the benchmark function optimization results for 100 dimensions can be appreciated. The results were averaged over 30 runs.

The third comparison is against Mahamed et al. [5], where they proposed a concept from swarm intelligence to enhance the performance of the HS. We use the same constrains used in [5], and in Tables 4.9 and 4.10 the results from [5] are presented.

From Table 4.9 the mean and standard deviation of the benchmark function optimization results for 30 dimensions can be appreciated, and only in the function six-hump camel back two dimensions are used. The results were averaged over 30 runs.

From Table 4.10 the mean and standard deviation of the benchmark function optimization results for 100 dimensions can be appreciated. The results were averaged

Table 4.9 Comparison against Mahamed et al. Results for 30 dimensions [5]

30 Dimensions					
GHS [5]			IT2-FHS HMR parameter		
Sphere	Mean	1.00E–05	Sphere	Mean	3.42E–10
	S.D.	2.20E–05		S.D.	3.64E–10
Rosenbrock	Mean	4.97E+01	Rosenbrock	Mean	4.54E–03
	S.D.	5.92E+01		S.D.	2.49E–02
Ackley	Mean	2.09E–02	Ackley	Mean	5.23E–07
	S.D.	2.17E–02		S.D.	4.57E–07
Griewank	Mean	1.02E–01	Griewank	Mean	4.60E–04
	S.D.	1.76E–01		S.D.	1.06E–03
Schwefel's	Mean	7.82e–02	Schwefel's	Mean	1.21e+00
	S.D.	1.14e–01		S.D.	1.01e–01
Rastrigin	Mean	8.63e–03	Rastrigin	Mean	4.29e–06
	S.D.	1.53e–02		S.D.	4.42e–06
Hyper-ellipsoid	Mean	5.15e+03	Hyper-ellipsoid	Mean	8.86e–06
	S.D.	6.35e+03		S.D.	8.61e–06
Six-hump camel back	Mean	−1.03e+00	Six-hump camel back	Mean	−1.03e+00
	S.D.	1.80e–05		S.D.	6.78e–16

Table 4.10 Comparison against Mahamed et al. Results for 100 dimensions [5]

100 Dimensions					
GHS [5]			IT2-FHS HMR parameter		
Sphere	Mean	2.23E+00	Sphere	Mean	3.23E–10
	S.D.	5.65E–01		S.D.	4.56E–10
Rosenbrock	Mean	2.60E+06	Rosenbrock	Mean	1.80E–03
	S.D.	9.16E+05		S.D.	9.86E–03
Ackley	Mean	8.77E+00	Ackley	Mean	1.02E–06
	S.D.	8.80E–01		S.D.	1.48E–06
Griewank	Mean	5.43E+01	Griewank	Mean	1.42E–02
	S.D.	1.86E+01		S.D.	3.78E–02
Schwefel's	Mean	1.90e+01	Schwefel's	Mean	3.62e+03
	S.D.	5.09e+00		S.D.	4.76e+02
Rastrigin	Mean	8.07e+01	Rastrigin	Mean	4.56e–06
	S.D.	3.04e+01		S.D.	7.22e–06
Hyper-ellipsoid	Mean	3.22e+05	Hyper-ellipsoid	Mean	7.01e–06
	S.D.	3.96e+04		S.D.	7.75e–06

over 30 runs. We can see that the more complex the problem of type-2 algorithm sometimes get better results.

4.1.1.3 Statistical Test

To validate the proposed method in the best way, it is decided to use the z-test test statistic, which is given by Eq. 4.4, and the parameters used for the test are shown in Table 4.11.

$$Z = \frac{\left(\bar{X}_1 - \bar{X}_2\right) - (\mu_1 - \mu_2)}{\sigma_{\bar{X}_1 - \bar{X}_2}} \tag{4.4}$$

The alternative hypothesis indicates that the IT2-FHS method is smaller than the global best HS (GHS) method and the null hypothesis indicates otherwise, with a rejection region for values less than -1.645. Table 4.12 shows the Z-values, "*S*" means

Table 4.11 Values for the statistical z-test

Parameter	Value
Level of confidence	95%
Alpha	0.05
H_a	$\mu_1 < \mu_2$ (Claim)
H_o	$\mu_1 \geq \mu_2$
Critical value	-1.645

Table 4.12 Results for the statistical test with FHS and HS

Function	Dimension	Z-value	Evidence
Rosenbrock	30	4.5979	S
	100	-15.5467	S
Ackley	30	-5.2752	S
	100	-54.5855	S
Griewank	30	-3.1599	S
	100	-15.9858	S
Sphere	30	-2.4869	S
	100	-21.6181	S
Schwefel's	30	40.7019	*N.S.*
	100	41.4335	*N.S.*
Rastrigin	30	-3.0879	S
	100	-14.5399	S
Rotated hyper-ellipsoid	30	-4.4422	S
	100	-44.537	S

that significant evidence was found and "*N.S.*" refers to the fact that no significant evidence was found.

The results presented in Table 4.12 show that for the interval type-2 fuzzy harmony search with 30 and 100 dimensions in some functions, there is significant evidence to reject the null hypothesis. It can be noted that by using the proposed method it is possible to obtain significant evidence in most of the mathematical benchmark functions, which indicates that by using the fuzzy system for the adjustment of parameters a control of the diversification and intensification in the search space is achieved, and in this better results are found. When applying the equation of the z-test, with a significance level of 0.05, the alternative hypothesis indicates the IT2-FHS method is smaller than the GHS method and the null hypothesis indicates otherwise, with a rejection region for values less than −1.645.

4.1.2   CEC 2015 Benchmark Mathematical Functions

For the experiments we used the set of 14 mathematical functions presented in Table 4.13, provided in the CEC 2015 competition [6–9]. In this case the problem is finding the global minimum for each mathematical function. These functions are tested with 10, 30, 50 and 100 dimensions.

Table 4.13 presents the equations used to perform the experiments; functions 1 and 2 are unimodal functions, 3 to 5 are simple multimodal functions, 6 to 8 are hybrid functions, 9 to 15 are composition functions, the search range used for all functions is [−100,100] and all experiments are executed 51 times and during each run, each function was evaluated for 10,000*D times. The parameter D denotes the dimension of the function. Table 4.14 presents the global minimum for each mathematical function.

Table 4.14 presents the local optimum, which is obtained using $F_i' = F_i - F_i*$ where F_i is the result obtained for the function and F_i* is the local optimum for each function and are used as g_i. In this way, the function values of the global optima of g_i are equal to 0 for all composition functions. A value of 100 was added to function 1, 200 to function 2, …, and a value of 1500 was added to function 15, all of which were subtracted from the calculated mean errors so that the best value for each function remains zero. With hybrid functions as the basic functions, the composition function can have different properties for different variables subcomponents.

4.1.2.1   Experiments and Results

The parameters used by the HS and the T1-FHS and IT2-FHS algorithms are indicated in Table 4.15.

Table 4.15 shows the parameter values used in each method, where we can find each value used as the number of harmonies, the dimensions and the maximum

Table 4.13 Mathematical functions for the experiments

No.	Equations
1	$f_1(x) = \sum_{i=1}^{D} (10^6)^{\frac{i-1}{D-1}} x_i^2$
2	$f_2(x) = x_1^2 + 10^6 \sum_{i=2}^{D} x_i^2$
3	$f_3(x) = 10^6 x_1^2 + \sum_{i=2}^{D} x_i^2$
4	$f_4(x) = \sum_{i=1}^{D-1} \left(100\left(x_i^2 - x_{i+1}\right)^2 + (x_i - 1)^2 \right)$
5	$f_5(x) = -20exp\left(-0.2\sqrt{\frac{1}{D}\sum_{i=1}^{D} x_i^2}\right) - exp\left(\frac{1}{D}\sum_{i=1}^{D}\cos(2\pi x_i)\right) + 20 + e$
6	$f_6(x) = \sum_{i=1}^{D}\left(\sum_{k=0}^{kmaz}\left[a^k \cos\left(2\pi b^k(x_i + 0.5)\right)\right]\right) - D\sum_{i=1}^{kmax}\left[a^k \cos 2\pi b^k \cdot 0.5\right]$ $$a = 0.5, \ b = 3, \ kmax = 20$$
7	$f_7(x) = \sum_{i=1}^{D}\frac{x_i^2}{4000} - \prod_{i=1}^{D}\cos\left(\frac{x_i}{\sqrt{i}}\right) + 1$
8	$f_8(x) = \sum_{i=1}^{D}\left(x_i^2 - 10\cos(2\pi x_i) + 10\right)$
9	$f_9(x) = 418.9829 \times D - \sum_{i=1}^{D} g(z_i), \quad z_i = x_i + 4.209687462275036e + 002$ $g(z_i) = z_i \sin\left(|z_i|^{\frac{1}{2}}\right) \ if |z_i| \leq 500$ $g(z_i) = (500 - mod(z_i, 500))\sin\left(\sqrt{|500 - mod(z_i, 500)|}\right) - \frac{(z_i - 500)^2}{10000D} \ if \ z_i > 500$ $g(z_i) = (mod(|z_i|, 500) - 500)sin\left(\sqrt{|mod(|z_i|, 500) - 500|}\right) - \frac{(z_i - 500)^2}{10000D} \ if \ z_i$
10	$f_{10}(x) = \frac{10}{D^2}\prod_{i=1}^{D}\left(1 + i\sum_{j=1}^{32}\frac{|2^j x_i - round(2^j x_i)|}{2^j}\right)^{\frac{10}{D^{1.2}}} - \frac{10}{D^2}$
11	$f_{11}(x) = \left|\sum_{i=1}^{D} x_i^2 - D\right|^{1/4} + \frac{\left(0.5\sum_{i=1}^{D} x_i^2 + \sum_{i=1}^{D} x_i\right)}{D+0.5}$
12	$f_{12}(x) = \left|\left(\sum_{i=1}^{D} x_i^2\right)^2 - \left(\sum_{i=1}^{D} x_i\right)^2\right|^{\frac{1}{2}} + \frac{\left(0.5\sum_{i=1}^{D} x_i^2 + \sum_{i=1}^{D} x_i\right)}{D+0.5}$
13	$f_{13}(x) = f_7(f_4(x_1, x_2)) + f_7(f_4(x_3, x_4)) + \ldots + f_7(f_4(x_{D-1}, x_D)) + f_7(f_4(x_D, x_1))$
14	Scaffer's F6 function: $g(x, y) = 0.5 + \frac{\left(\sin^2\left(\sqrt{x^2+y^2}\right) - 0.5\right)}{(1+0.001(x^2+y^2))^2}$ $f_{14}(x) = g(x_1, x_2) + g(x_2, x_3) + \ldots + g(x_{D-1}, x_D) + g(x_D, x_1)$

Table 4.14 Global minimum for each function

No.	Functions	$F_i^* = F_i(x^*)$
F1	Rotated high-conditioned elliptic function	100
F2	Rotated cigar function	200
F3	Shifted and rotated Ackley's function	300
F4	Shifted and rotated Rastrigin's function	400
F5	Shifted and rotated Schwefel's function	500
F6	Hybrid function 1 ($N = 3$)	600
F7	Hybrid function 2 ($N = 4$)	700
F8	Hybrid function 3 ($N = 5$)	800
F9	Composition function 1 ($N = 3$)	900
F10	Composition function 2 ($N = 3$)	1000
F11	Composition function 3 ($N = 5$)	1100
F12	Composition function 4 ($N = 5$)	1200
F13	Composition function 5 ($N = 5$)	1300
F14	Composition function 6 ($N = 7$)	1400
F15	Composition function 7 ($N = 10$)	1500

Table 4.15 Parameter values used in the methods

Parameter	Simple HS	T1-FHS and IT2-FHS
Harmonies	100	100
Dimensions	10,30,50,100	10, 30, 50, 100
Iterations	100,000, 300,000, 500,000, 1,000,000	100,000, 300,000, 500,000, 1,000,000
HMR	0.95	Dynamic
PArate	0.75	0.75

number of iterations for each dimension number and value of the *HMR* and *PArate* parameters used in each method.

In this case, 51 experiments were performed for each mathematical function using the HS and the T1-FHS and IT2-FHS methods. Tables 4.16 and 4.19 present the results of the experiments when the T1-FHS was applied to the 15 benchmark functions with 10, 30, 50 and 100 dimensions.

Tables 4.16 and 4.17 show the averages and standard deviations (S.D) obtained from the 51 runs applied to the F1 to F15 functions using the original harmony search algorithm (HS) and the fuzzy harmony search algorithm (T1-FHS) method with 10 and 30 dimensions, respectively.

Tables 4.18 and 4.19 show the averages and standard deviations obtained from the 51 runs applied to the functions F1 to the F15 using the original harmony search

Table 4.16 Summary of results obtained with 10 dimensions with the HS and T1-FHS methods

Function	10 Dimensions			
	HS		T1-FHS	
	Mean	S.D.	Mean	S.D.
F1	1.91E+04	2.47E+05	2.62E–02	1.58E–02
F2	7.10E+02	9.31E+03	1.51E+00	4.63E–01
F3	3.20E+02	2.84E+00	2.05E+01	2.94E+00
F4	4.08E+02	3.82E+00	1.24E+02	2.52E+01
F5	7.25E+02	1.79E+02	2.43E+03	4.24E+02
F6	1.32E+03	1.07E+03	4.05E–03	2.83E–03
F7	7.01E+02	6.39E–01	5.70E+01	2.70E+01
F8	1.44E+03	1.69E+03	3.61E–04	2.57E–04
F9	1.08E+03	2.98E+01	1.00E+02	4.68E–02
F10	5.73E+05	2.94E+06	1.11E+03	1.05E+03
F11	1.70E+03	1.32E+02	2.63E+02	1.06E+02
F12	1.36E+03	1.05E+01	1.02E+02	6.19E–01
F13	1.39E+03	5.02E+01	3.20E+01	2.53E+00
F14	1.98E+04	2.97E+03	5.44E+03	2.81E+03
F15	3.53E+03	2.62E+03	1.00E+02	0.00E+00

Table 4.17 Summary of results obtained with 30 dimensions with the HS and T1-FHS methods

Function	30 Dimensions			
	HS		T1-FHS	
	Mean	S.D.	Mean	S.D.
F1	1.91E+04	2.47E+05	2.62E–02	1.58E–02
F2	7.10E+02	9.31E+03	1.51E+00	4.63E–01
F3	3.20E+02	2.84E+00	2.05E+01	2.94E+00
F4	4.08E+02	3.82E+00	1.24E+02	2.52E+01
F5	7.25E+02	1.79E+02	2.43E+03	4.24E+02
F6	1.32E+03	1.07E+03	4.05E–03	2.83E–03
F7	7.01E+02	6.39E–01	5.70E+01	2.70E+01
F8	1.44E+03	1.69E+03	3.61E–04	2.57E–04
F9	1.08E+03	2.98E+01	1.00E+02	4.68E–02
F10	5.73E+05	2.94E+06	1.11E+03	1.05E+03
F11	1.70E+03	1.32E+02	2.63E+02	1.06E+02
F12	1.36E+03	1.05E+01	1.02E+02	6.19E–01
F13	1.39E+03	5.02E+01	3.20E+01	2.53E+00
F14	1.98E+04	2.97E+03	5.44E+03	2.81E+03
F15	3.53E+03	2.62E+03	1.00E+02	0.00E+00

Table 4.18 Summary of results obtained with 50 dimensions with the HS and T1-FHS methods

Function	50 Dimensions			
	HS		T1-FHS	
	Mean	S.D.	Mean	S.D.
F1	1.91E+04	2.47E+05	2.62E–02	1.58E–02
F2	7.10E+02	9.31E+03	1.51E+00	4.63E–01
F3	3.20E+02	2.84E+00	2.05E+01	2.94E+00
F4	4.08E+02	3.82E+00	1.24E+02	2.52E+01
F5	7.25E+02	1.79E+02	2.43E+03	4.24E+02
F6	1.32E+03	1.07E+03	4.05E–03	2.83E–03
F7	7.01E+02	6.39E–01	5.70E+01	2.70E+01
F8	1.44E+03	1.69E+03	3.61E–04	2.57E–04
F9	1.08E+03	2.98E+01	1.00E+02	4.68E–02
F10	5.73E+05	2.94E+06	1.11E+03	1.05E+03
F11	1.70E+03	1.32E+02	2.63E+02	1.06E+02
F12	1.36E+03	1.05E+01	1.02E+02	6.19E–01
F13	1.39E+03	5.02E+01	3.20E+01	2.53E+00
F14	1.98E+04	2.97E+03	5.44E+03	2.81E+03
F15	3.53E+03	2.62E+03	1.00E+02	0.00E+00

Table 4.19 Summary of results obtained with 100 dimensions with the HS and T1-FHS methods

Function	100 Dimensions			
	HS		T1-FHS	
	Mean	S.D.	Mean	S.D.
F1	1.91E+04	2.47E+05	2.62E–02	1.58E–02
F2	7.10E+02	9.31E+03	1.51E+00	4.63E–01
F3	3.20E+02	2.84E+00	2.05E+01	2.94E+00
F4	4.08E+02	3.82E+00	1.24E+02	2.52E+01
F5	7.25E+02	1.79E+02	2.43E+03	4.24E+02
F6	1.32E+03	1.07E+03	4.05E–03	2.83E–03
F7	7.01E+02	6.39E–01	5.70E+01	2.70E+01
F8	1.44E+03	1.69E+03	3.61E–04	2.57E–04
F9	1.08E+03	2.98E+01	1.00E+02	4.68E–02
F10	5.73E+05	2.94E+06	1.11E+03	1.05E+03
F11	1.70E+03	1.32E+02	2.63E+02	1.06E+02
F12	1.36E+03	1.05E+01	1.02E+02	6.19E–01
F13	1.39E+03	5.02E+01	3.20E+01	2.53E+00
F14	1.98E+04	2.97E+03	5.44E+03	2.81E+03
F15	3.53E+03	2.62E+03	1.00E+02	0.00E+00

Table 4.20 Results obtained with 10 dimensions using the IT2-FHS method

Function	Best	Worst	Median	Mean	S.D.
1	**7.10E–03**	**7.52E–01**	**3.28E–02**	**6.37E–02**	**1.25E–01**
2	2.07E+09	2.50E+10	1.60E+10	1.64E+10	5.54E+09
3	3.20E+02	3.21E+02	3.21E+02	3.21E+02	1.86E–01
4	4.74E+02	5.77E+02	5.32E+02	5.32E+02	2.14E+01
5	2.43E+03	3.45E+03	3.01E+03	2.99E+03	2.23E+02
6	**1.00E–04**	**5.10E–02**	**1.60E–03**	**4.28E–03**	**9.14E–03**
7	7.09E+02	8.58E+02	7.54E+02	7.61E+02	3.38E+01
8	**0.00E+00**	**4.00E–03**	**3.00E–04**	**5.67E–04**	**7.07E–04**
9	1.03E+03	1.16E+03	1.07E+03	1.08E+03	2.64E+01
10	**0.00E+00**	**9.40E–03**	**3.00E–04**	**9.43E–04**	**1.73E–03**
11	1.47E+03	1.90E+03	1.64E+03	1.66E+03	1.18E+02
12	1.33E+03	1.38E+03	1.36E+03	1.36E+03	1.19E+01
13	1.35E+03	1.64E+03	1.39E+03	1.40E+03	5.01E+01
14	1.13E+04	2.67E+04	2.07E+04	2.10E+04	3.04E+03
15	1.66E+03	1.23E+04	4.09E+03	5.12E+03	2.99E+03

algorithm (HS) and the type-1 fuzzy harmony search algorithm (T1-FHS) method with 50 and 100 dimensions, respectively.

From the previous tables it can be noted that on average the proposed T1-FHS has better performance than the original HS algorithm in most of the functions.

Tables 4.20 and 4.22 present the results of the experiments when IT2-FHS applied to the 15 benchmark functions with 10, 30 and 50 dimensions. Table (4.21).

It can be seen that by using the IT2-FHS to the 15 reference functions, good results are obtained for functions 1, 6, 8 and 10.

4.1.2.2 Statistical Test

To statistically verify the efficiency of the proposed method, it is decided to use the z-test test statistic, which is given by Eq. 4.4, and the parameters used for the test are shown in Table 4.23.

Table 4.21 Results obtained with 30 dimensions using the IT2-FHS method

Function	Best	Worst	Median	Mean	S.D.
1	**1.76E–02**	**3.91E–01**	**4.43E–02**	**5.08E–02**	**5.03E–02**
2	1.04E+00	1.55E+11	1.48E+00	3.03E+09	2.16E+10
3	2.11E+01	3.21E+02	2.13E+01	2.72E+01	4.20E+01
4	5.14E+02	1.09E+03	6.59E+02	6.49E+02	8.53E+01
5	7.96E+03	1.02E+04	9.33E+03	9.38E+03	4.49E+02
6	**4.00E–04**	**2.92E+08**	**3.50E–03**	**5.72E+06**	**4.08E+07**
7	6.61E+02	3.03E+03	1.31E+03	1.35E+03	5.16E+02
8	**1.00E–04**	**6.79E+07**	**1.30E–03**	**1.60E+06**	**9.66E+06**
9	**1.00E–04**	**1.03E+03**	**7.55E+02**	**7.36E+02**	**1.49E+02**
10	**6.00E–04**	**1.96E+08**	**3.50E–03**	**3.85E+06**	**2.75E+07**
11	1.80E+03	3.25E+03	2.26E+03	2.36E+03	3.85E+02
12	1.96E+02	1.46E+03	2.60E+02	2.84E+02	1.69E+02
13	2.30E+02	3.23E+03	1.16E+03	1.31E+03	7.66E+02
14	8.57E+04	2.64E+05	1.56E+05	1.59E+05	3.78E+04
15	1.00E–04	1.16E+06	4.95E+05	5.12E+05	2.67E+05

Table 4.22 Results obtained with 50 dimensions using the IT2-FHS method

Function	Best	Worst	Median	Mean	S.D.
1	**4.72E–02**	**2.01E–01**	**1.11E–01**	**1.14E–01**	**3.09E–02**
2	2.10E+11	3.37E+11	2.82E+11	2.81E+11	2.96E+10
3	2.13E+01	2.15E+01	2.15E+01	2.14E+01	4.60E–02
4	1.05E+03	1.40E+03	1.22E+03	1.22E+03	7.88E+01
5	1.51E+04	1.75E+04	1.67E+04	1.65E+04	5.56E+02
6	**4.90E–03**	**1.95E+09**	**7.34E+08**	**7.24E+08**	**4.49E+08**
7	1.69E+03	7.80E+03	4.38E+03	4.33E+03	1.49E+03
8	**1.20E–03**	**8.70E–03**	**3.50E–03**	**3.68E–03**	**1.47E–03**
9	9.30E+02	1.93E+03	1.54E+03	1.52E+03	2.33E+02
10	**3.20E–03**	**1.83E–02**	**9.30E–03**	**9.30E–03**	**3.45E–03**
11	2.91E+03	5.32E+03	3.61E+03	3.70E+03	5.56E+02
12	3.23E+02	4.98E+02	4.14E+02	4.08E+02	3.74E+01
13	1.11E+03	7.91E+03	3.80E+03	3.65E+03	1.30E+03
14	2.77E+05	8.48E+05	5.25E+05	5.48E+05	1.60E+05
15	9.69E+05	3.15E+06	1.62E+06	1.76E+06	5.73E+05

Table 4.23 Values for the statistical z-test

Parameter	Value
Level of confidence	95%
Alpha	0.05
H_a	$\mu_1 < \mu_2$ (Claim)
H_o	$\mu_1 \geq \mu_2$
Critical value	−1.645

The alternative hypothesis indicates that the T1-FHS method is smaller than the HS method and the null hypothesis indicates otherwise, with a rejection region for values less than −1.645. Table 4.24 shows the Z-values, "S" means that significant evidence was found and "N.S." refers to the fact that no significant evidence was found.

The results presented in Table 4.24 are that: for the type-1 fuzzy harmony search with 10, 30, 50 and 100 dimensions in some functions, there is significant evidence to reject the null hypothesis. It can be noted that by using the proposed method it is possible to obtain significant evidence in most of the mathematical benchmark functions, which indicates that by using the type-1 fuzzy system for the adjustment of parameters a control of the diversification and intensification in the search space is achieved, and in this better results are found. When applying the equation of the z-test, with a significance level of 0.05, the alternative hypothesis indicates that the T1-FHS method is smaller than the HS method and the null hypothesis indicates otherwise, with a rejection region for values less than −1.645.

Table 4.24 Results for the statistical test with the T1-FHS and HS methods

Function	10 Dimensions		30 Dimensions		50 Dimensions		100 Dimensions	
	Z-value	Evidence	Z-value	Evidence	Z-value	Evidence	Z-value	Evidence
F1	−0.55	N.S.	−3.84	S	−8.62	S	−9.80	S
F2	−0.54	N.S.	−0.07	N.S.	−0.005	N.S.	−0.003	N.S.
F3	−523	S	−716.47	S	−716.61	S	−716	S
F4	−8.04	S	12.34	N.S.	7.58	N.S.	−5.153	S
F5	26.45	N.S.	52.19	N.S.	13.84	N.S.	−9.94	S
F6	−8.81	S	−3.32	S	−6.03	S	−11.40	S
F7	−170	S	10.39	N.S.	8.92	N.S.	7.27	N.S.
F8	−6.08	S	−6.21	S	−3.49	S	−6.40	S
F9	−234	S	−70.30	S	−68.22	S	−41.84	S
F10	−1.38	N.S.	−5.45	S	−17.18	S	−0.02	N.S.
F11	−60.61	S	−63.01	S	−46.06	S	−27.12	S
F12	−854	S	−337	S	−285	S	−97.90	S
F13	−196	S	−19	S	−24.01	S	−13.54	S
F14	−25.08	S	−22.63	S	−19.41	S	−15.27	S
F15	−9.34	S	−10.43	S	−19.81	S	−0.24	N.S.

4.1.3   CEC 2017 Benchmark Mathematical Functions

The 30 mathematical functions presented in Table 4.25 are provided in the CEC 2017 competition. In this case the problem is to find the global minimum for each mathematical function. These functions are tested with 10, 30, 50 and 100 dimensions.

For all functions the global minimum is 0, as can be appreciated in Table 4.25.

4.1.3.1   Proposed Methodology

In Sect. 4.1.1.1, we presented the proposed method which was composed of one input and one output. For this case of study, it is decided to add a second input to the fuzzy system which is called diversity, whose function is to measure the dispersion of the harmonies and is calculated by the Euclidean distance, as shown in Eq. 4.5.

$$\text{Diversity (HM(t))} = \frac{1}{n_{HM}} \sum_{i=1}^{n_{HM}} \sqrt{\sum_{j=1}^{n_x} \left(x_{ij}(t) - \bar{x}_j(t) \right)^2} \tag{4.5}$$

Figure 4.5 shows the complete fuzzy system.

The design of the input and output variables can be appreciated in Fig. 4.5; these are granulated into three triangular membership functions. The linguistic values are low, medium and high. The rules were created based on the function of each parameter and are defined in Fig. 4.6.

4.1.3.2   Experiments and Results

A total of 30 benchmark mathematical functions of the CEC 2017 were used for the experimentation of the original and fuzzy methods. Dimensions of 10, 30, 50, 100 variables were used with 51 runs for each function to achieve a comparison with the methods presented in this section.

The parameters used in each method can be appreciated in Table 4.26.

Table 4.26 shows the parameters used in each method, which are harmony memory (HMS), pitch adjustment (*PArate*), harmony memory accepting (*HMR*) and dimensions.

In the study 51 experiments were carried out and in Table 4.27 the results of the average obtained by each mathematical function are shown using 10 and 30 dimensions for each method.

In Table 4.28 the results of the average obtained by each mathematical function are shown using 50 and 100 dimensions for each method.

Table 4.25 CEC 2017 benchmark functions

	No.	Functions	$F_i^* = F_i(x^*)$
Unimodal functions	1	Shifted and rotated bent cigar function	100
	2	Shifted and rotated sum of different power function	200
	3	Shifted and rotated Zakharov function	300
Simple multimodal functions	4	Shifted and rotated Rosenbrock's function	400
	5	Shifted and rotated Rastrigin's function	500
	6	Shifted and rotated expanded Scaffer's F6 function	600
	7	Shifted and rotated Lunacek Bi_Rastrigin function	700
	8	Shifted and rotated non-continuous Rastrigin's function	800
	9	Shifted and rotated Levy function	900
	10	Shifted and rotated Schwefel's function	1000
Hybrid functions	11	Hybrid function 1 (N=3)	1100
	12	Hybrid function 1 (N=3)	1200
	13	Hybrid function 2 (N=3)	1300
	14	Hybrid function 3 (N=3)	1400
	15	Hybrid function 4(N=4)	1500
	16	Hybrid function 5 (N=4)	1600
	17	Hybrid function 6 (N=4)	1700
	18	Hybrid function 6 (N=5)	1800
	19	Hybrid function 6 (N=5)	1900
	20	Hybrid function 6 (N=5)	2000
	21	Composition function 1 (N=3)	2100
	22	Composition function 2 (N=3)	2200
	23	Composition function 3 (N=4)	2300
	24	Composition function 4 (N=4)	2400
	25	Composition function 5 (N=5)	2500
	26	Composition function 6 (N=5)	2600
	27	Composition function 7 (N=6)	2700
	28	Composition function 8 (N=6)	2800
	29	Composition function 9 (N=3)	2900
	30	Composition function 10 (N=3)	3000

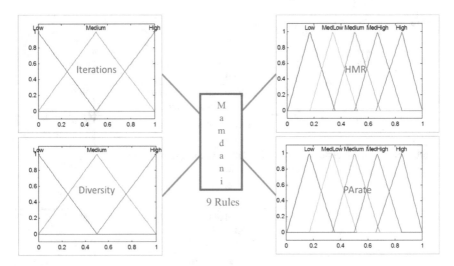

Fig. 4.5 Complete fuzzy system

1. If (Iteration is Low) and (D is Low) then (HMR is High) (PArate is Low)
2. If (Iteration is Low) and (D is Medium) then (HMR is Medium) (PArate is Medium)
3. If (Iteration is Low) and (D is High) then (HMR is Medium) (PArate is Medium)
4. If (Iteration is Medium) and (D is Low) then (HMR is Medium) (PArate is Medium)
5. If (Iteration is Medium) and (D is Medium) then (HMR is Medium) (PArate is Medium)
6. If (Iteration is Medium) and (D is High) then (HMR is Medium) (PArate is Medium)
7. If (Iteration is High) and (D is Low) then (HMR is Medium) (PArate is High)
8. If (Iteration is High) and (D is Medium) then (HMR is Medium) (PArate is Medium)
9. If (Iteration is High) and (D is High) then (HMR is Low) (PArate is High)

Fig. 4.6 Rules for the fuzzy system FHS

Table 4.26 Parameters used for test problems

Methods	HMS	PArate	HMR	Dimensions
Simple HS	40	0.75	0.95	10, 30, 50, 100
Fuzzy HS	40	Dynamic	Dynamic	10, 30, 50, 100

4.1.3.3   Statistical Test

A z-test statistic was conducted to validate the effectiveness of the proposed method, which is given by Eq. 4.4, and the parameters used for the test are shown in Table 4.29.

The alternative hypothesis indicates that the T1-FHS method is smaller than the HS method and the null hypothesis indicates otherwise, with a rejection region for values less than −1.645. Table 4.30 shows the Z-values, "S" means that significant

Table 4.27 Results for 10 and 30 dimensions

Function	HS		FHS		HS		FHS	
	10		10		30		30	
	Mean	S.D.	Mean	S.D.	Mean	S.D.	Mean	S.D.
1	**2.54E+03**	2.54E+03	2.72E+03	2.62E+03	**4.02E+03**	4.61E+03	5.95E+03	5.92E+03
2	1.96E−02	1.40E−01	**0.00E+00**	0.00E+00	**1.96E−02**	1.40E−01	5.39E+13	3.56E+14
3	1.18E−08	3.35E−08	**0.00E+00**	0.00E+00	**1.96E+01**	2.16E+01	1.94E+01	2.03E+01
4	3.67E−01	8.97E−01	2.12E+00	6.92E−01	**1.19E+02**	2.48E+01	1.24E+02	2.58E+01
5	8.04E+00	3.43E+00	4.23E+00	1.96E+00	**5.56E+01**	1.41E+01	5.52E+01	1.65E+01
6	2.79E−08	1.99E−07	**0.00E+00**	0.00E+00	1.29E−02	7.00E−02	**8.94E−04**	3.16E−03
7	1.88E+01	7.35E+00	**1.26E+01**	4.63E+00	1.01E+02	2.52E+01	**1.01E+02**	2.73E+01
8	9.15E+00	4.04E+00	**4.37E+00**	1.77E+00	6.30E+01	1.51E+01	**6.09E+01**	1.43E+01
9	0.00E+00	0.00E+00	**0.00E+00**	0.00E+00	3.75E+01	5.03E+01	**3.56E+01**	4.37E+01
10	3.03E+02	2.14E+02	**1.86E+02**	1.02E+02	**2.49E+03**	6.61E+02	2.53E+03	6.94E+02
11	4.76E+00	2.41E+00	**3.08E+00**	1.58E+00	**5.57E+01**	3.48E+01	6.20E+01	3.70E+01
12	2.12E+04	1.68E+04	**1.54E+04**	1.63E+04	2.57E+05	4.63E+05	**2.46E+05**	3.95E+05
13	7.32E+03	8.03E+03	**1.25E+03**	2.11E+03	1.11E+04	1.52E+04	**9.88E+03**	1.42E+04
14	2.68E+01	1.23E+01	**8.28E+00**	8.44E+00	8.04E+03	6.40E+03	**5.64E+03**	4.60E+03
15	3.39E+01	3.47E+01	**1.73E+00**	1.44E+00	**1.06E+04**	8.73E+03	1.61E+04	9.12E+03
16	1.77E+01	4.30E+01	**3.06E−01**	1.89E−01	**5.16E+02**	2.65E+02	5.79E+02	2.05E+02
17	1.38E+01	1.25E+01	**4.78E+00**	7.65E+00	1.58E+02	9.03E+01	**1.55E+02**	8.84E+01
18	4.55E+03	4.10E+03	**7.43E+02**	1.14E+03	**2.23E+05**	2.04E+05	2.90E+05	2.36E+05
19	2.39E+01	2.55E+01	**1.56E+00**	1.25E+00	2.00E+04	1.72E+04	**1.89E+04**	1.57E+04

(continued)

Table 4.27 (continued)

Function	HS		FHS		HS		FHS	
	10		10		30		30	
	Mean	S.D.	Mean	S.D.	Mean	S.D.	Mean	S.D.
20	9.63E+00	1.01E+01	**5.31E+00**	2.27E+01	**1.42E+02**	9.21E+01	1.51E+02	9.21E+01
21	**1.82E+02**	5.05E+01	1.87E+02	4.34E+01	**2.63E+02**	1.59E+01	2.64E+02	1.38E+01
22	9.64E+01	2.34E+01	**9.65E+01**	2.25E+01	1.93E+03	1.36E+03	**1.91E+03**	1.39E+03
23	3.08E+02	4.17E+00	**3.06E+02**	2.41E+00	4.03E+02	1.18E+01	**3.99E+02**	1.22E+01
24	3.25E+02	5.70E+01	**3.20E+02**	6.50E+01	**4.82E+02**	2.68E+01	4.83E+02	3.51E+01
25	4.15E+02	2.30E+01	**4.07E+02**	1.79E+01	4.03E+02	2.23E+01	**3.97E+02**	1.23E+01
26	3.48E+02	2.33E+02	**3.19E+02**	1.37E+02	1.62E+03	2.04E+02	**1.44E+03**	4.59E+02
27	3.98E+02	1.40E+01	**3.93E+02**	6.99E+00	5.25E+02	1.24E+01	**5.25E+02**	1.02E+01
28	4.22E+02	1.56E+02	**3.94E+02**	1.48E+02	4.22E+02	4.36E+01	**4.20E+02**	3.84E+01
29	2.62E+02	3.31E+01	**2.37E+02**	1.59E+01	5.95E+02	1.04E+02	**5.80E+02**	1.17E+02
30	5.59E+05	8.94E+05	**2.06E+05**	5.65E+05	**8.62E+03**	3.64E+03	9.47E+03	3.92E+03

Table 4.28 Results for 50 and 100 dimensions

Function	HS 50		FHS 50		HS 100		FHS 100	
	Mean	S.D.	Mean	S.D.	Mean	S.D.	Mean	S.D.
1	1.62E+07	1.04E+08	**7.74E+04**	4.04E+05	7.84E+08	3.24E+08	**3.81E+06**	2.13E+07
2	**1.96E−02**	1.40E−01	2.56E+36	1.63E+37	5.25E+119	2.93E+120	**2.38E+102**	1.70E+103
3	1.16E+05	2.45E+04	**1.22E+04**	2.66E+03	5.06E+05	5.38E+04	**1.64E+05**	2.46E+04
4	3.35E+02	7.95E+01	**2.49E+02**	4.96E+01	8.73E+02	2.16E+02	**5.63E+02**	9.30E+01
5	1.50E+02	3.50E+01	**1.32E+02**	2.18E+01	5.98E+02	1.52E+02	**3.53E+02**	5.90E+01
6	3.99E+00	2.03E+00	**2.60E−01**	2.75E−01	2.17E+01	5.13E+00	**8.89E+00**	2.65E+00
7	3.23E+02	7.73E+01	**2.44E+02**	4.11E+01	1.20E+03	1.74E+02	**8.57E+02**	1.13E+02
8	1.49E+02	3.81E+01	**1.33E+02**	2.73E+01	5.86E+02	1.39E+02	**3.52E+02**	5.31E+01
9	1.46E+03	1.46E+03	**5.43E+02**	3.65E+02	1.48E+04	5.58E+03	**7.01E+03**	8.63E+03
10	9.01E+03	2.56E+03	**5.51E+03**	1.14E+03	2.69E+04	3.10E+03	**1.57E+04**	5.35E+03
11	3.65E+02	1.13E+02	**2.21E+02**	7.02E+01	2.80E+04	7.74E+03	**1.54E+03**	3.58E+02
12	1.65E+07	3.80E+07	**3.23E+06**	3.27E+06	3.68E+08	3.09E+08	**6.35E+07**	7.30E+07
13	**8.10E+03**	8.00E+03	8.52E+03	8.43E+03	3.75E+05	2.60E+06	**4.84E+04**	2.64E+05
14	2.08E+05	1.64E+05	**7.89E+04**	6.14E+04	4.99E+06	3.66E+06	**2.04E+06**	1.75E+06
15	4.98E+04	3.07E+05	**1.07E+04**	9.69E+03	1.96E+06	8.46E+06	**1.88E+04**	5.69E+04
16	1.51E+03	3.99E+02	**1.31E+03**	3.40E+02	5.12E+03	1.66E+03	**3.40E+03**	6.59E+02
17	1.25E+03	2.99E+02	**1.03E+03**	3.18E+02	3.57E+03	7.53E+02	**2.91E+03**	5.77E+02
18	3.29E+06	2.87E+06	**1.52E+06**	1.34E+06	1.10E+07	5.07E+06	**4.56E+06**	2.78E+06
19	1.29E+04	1.45E+04	**1.11E+04**	1.27E+04	1.39E+06	9.91E+06	**6.63E+03**	7.06E+03

(continued)

Table 4.28 (continued)

Function	HS 50		FHS 50		HS 100		FHS 100	
	Mean	S.D.	Mean	S.D.	Mean	S.D.	Mean	S.D.
20	9.51E+02	2.75E+02	6.00E+02	2.65E+02	4.40E+03	9.76E+02	3.09E+03	8.35E+02
21	3.65E+02	5.31E+01	3.32E+02	2.83E+01	8.39E+02	1.30E+02	6.14E+02	5.34E+01
22	9.79E+03	2.31E+03	6.04E+03	1.69E+03	2.84E+04	3.11E+03	1.65E+04	3.52E+03
23	6.21E+02	7.43E+01	5.62E+02	3.40E+01	9.97E+02	8.68E+01	8.28E+02	4.28E+01
24	8.44E+02	9.20E+01	6.58E+02	7.93E+01	1.71E+03	1.88E+02	1.28E+03	5.92E+01
25	6.62E+02	3.89E+01	6.14E+02	3.60E+01	1.70E+03	2.85E+02	1.22E+03	1.44E+02
26	2.82E+03	5.15E+02	2.59E+03	3.70E+02	9.32E+03	1.33E+03	7.31E+03	7.02E+02
27	7.94E+02	7.42E+01	6.61E+02	5.77E+01	1.12E+03	1.00E+02	8.88E+02	5.38E+01
28	5.87E+02	4.11E+01	5.30E+02	4.29E+01	1.55E+03	4.65E+02	9.10E+02	8.17E+01
29	1.20E+03	3.54E+02	8.08E+02	2.16E+02	4.55E+03	7.40E+02	3.26E+03	5.04E+02
30	3.44E+06	9.86E+05	2.15E+06	6.65E+05	3.18E+05	3.42E+05	3.11E+04	2.50E+04

Table 4.29 Values for the statistical z-test

Parameter	Value
Level of confidence	95%
Alpha	0.05
H_a	$\mu_1 < \mu_{2 \text{ (Claim)}}$
H_o	$\mu_1 \geq \mu_2$
Critical value	-1.645

Table 4.30 Results for the statistical test with T1-FHS and HS

Function	10 Dimensions		30 Dimensions		50 Dimensions		100 Dimensions	
	Z-value	Evidence	Z-value	Evidence	Z-value	Evidence	Z-value	Evidence
F1	0.3435	N.S.	1.83	N.S.	−1.09	N.S.	−17.20	S
F2	−1	N.S.	1.08	N.S.	1.12	N.S.	−1.27	N.S.
F3	−2.51	S	−0.05	N.S.	−30.12	S	−26.13	S
F4	−9.75	S	1.14	N.S.	−6.58	S	−8.05	S
F5	−6.88	S	−0.13	N.S.	−3.27	S	−9.26	S
F6	−1	N.S.	−1.22	N.S.	−12.98	S	−13.80	S
F7	−5.13	S	0	N.S.	−6.42	S	−8.76	S
F8	−7.71	S	−0.70	N.S.	−2.58	S	−9.50	S
F9	0	N.S.	−0.20	N.S.	−4.37	S	−5.12	S
F10	−3.53	S	0.30	N.S.	−8.93	S	−10.62	S
F11	−4.16	S	0.89	N.S.	−7.70	S	−21.73	S
F12	−1.77	S	−0.13	N.S.	−2.49	S	−6.72	S
F13	−5.22	S	−0.41	N.S.	0.25	N.S.	−0.88	N.S.
F14	−8.84	S	−2.16	S	−5.27	S	−5.03	S
F15	−6.61	S	3.11	N.S.	−0.90	N.S.	−1.63	N.S.
F16	−2.88	N.S.	1.35	N.S.	−2.72	S	−6.12	S
F17	−4.37	S	−0.17	N.S.	−3.49	S	−3.96	S
F18	−6.39	S	1.54	N.S.	−3.97	S	−7.56	S
F19	−6.25	S	−0.33	N.S.	−0.64	N.S.	−1.0	N.S.
F20	−1.23	N.S.	0.45	N.S.	−6.56	S	−6.21	S
F21	0.51	N.S.	0.21	N.S.	−3.84	S	−8.27	S
F22	0.02	N.S.	−0.07	N.S.	−9.33	S	−13.41	S
F23	−3.17	S	−1.70	S	−5.11	S	−6.39	S
F24	−0.39	N.S.	0.23	N.S.	−10.94	S	−9.43	S
F25	−2.16	S	−16.13	S	−6.52	S	−8.10	S
F26	−0.75	N.S.	−2.47	S	−2.55	S	−6.66	S
F27	−2.38	S	−0.08	N.S.	−10.08	S	−7.90	S
F28	−0.92	N.S.	−0.26	N.S.	−6.76	S	−8.48	S
F29	−4.87	S	−0.69	N.S.	−6.78	S	−7.92	S
F30	−2.38	S	1.14	N.S.	−7.70	S	−5.92	S

evidence was found and "*N.S.*" refers to the fact that no significant evidence was found.

The results presented in Table 4.30 are that: for the type-1 fuzzy harmony search with 10, 30, 50 and 100 dimensions in some functions, there is significant evidence to reject the null hypothesis. It can be noted that by using the proposed method it is possible to obtain significant evidence in some of the mathematical benchmark functions, which it indicates that by using the type-1 fuzzy system for the adjustment of parameters a control of the diversification and intensification in the search space is achieved, and in this better results are found. When applying the equation of the z-test, with a significance level of 0.05, the alternative hypothesis indicates the T1-FHS method is smaller than the HS method and the null hypothesis indicates otherwise, with a rejection region for values less than -1.645.

4.2   Benchmark Control Problems

In the literature there are several models of fuzzy control systems, one of the most used is the Mamdani model, and the other is the Takagi-Sugeno-Kang (TSK) model. To validate the proposed algorithms in this chapter, various benchmark control problems were selected that use a Mamdani model and that uses a TSK model.

4.2.1   Mamdani Controller

The Mamdani model works with the main sections of a fuzzy logic system, which are the fuzzifier that takes crisp inputs and maps them into fuzzy sets, the inference engine based on the rules, maps fuzzy sets from the antecedents to fuzzy sets from the consequents; finally, the defuzzifier obtains a crisp value from the fuzzy sets.

4.2.1.1   Water Tank Controller

The first benchmark control problem is the water level control in a tank; the main goal of the water tank controller is to maintain the level of water in a tank and the activation of the valve to keep the tank full, preventing the water from draining out of the tank. The diagram of the operation of this problem is shown in Fig. 4.7.

The mathematical model of this controller is expressed in Eq. 4.6:

$$h(t) = h(0) + \int_0^t \frac{1}{A}\left(q_{in}(t') - q_{out}(t')\right)dt' = \int_0^t \frac{1}{A}\left(q_{in}(t') - a\sqrt{2gh(t')}\right)dt'$$

$$(4.6)$$

Fig. 4.7 Water tank system

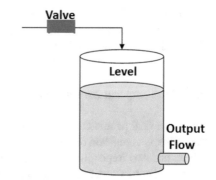

The fuzzy system for this problem is of Mamdani type, which is composed of two inputs which have three triangular membership functions and one output which has five triangular membership functions, as represented in Fig. 4.8, and the fuzzy rules that are used are shown in Fig. 4.9.

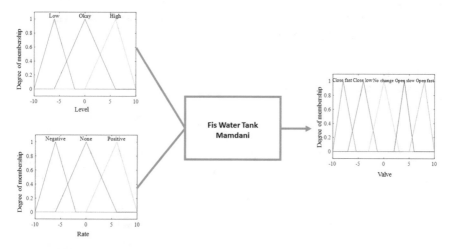

Fig. 4.8 Water tank system

1. If (Level is Okay) then (Valve is no_change)
2. If (Level is High) then (Valve is open_fast)
3. If (Level is Low) then (Valve is close_fast)
4. If (Level is Okay) then (Valve is Positive) then (valve is close_slow)
5. If (Level is Okay) then (Valve is Negative) then (valve is open_slow)

Fig. 4.9 Fuzzy rules for the water tank controller

The fuzzy rules that are outlined in Fig. 4.9 are those that maintain control of the opening of the valve and with this the level of water in the tank. These rules were obtained based on analyzing the filling of a water tank.

4.2.1.2 Temperature Controller

The main goal of this controller is to regulate the temperature and water flow. This controller is composed of two inputs and two outputs, and the first input is the temp, which has two trapezoidal and one triangular membership functions with linguistic variables of Cold, Good and Hot. The second input is the flow, which has two trapezoidal and one triangular membership functions with linguistic variables of Soft, Good and Hard. The first output is the cold variable which has the five triangular membership functions with linguistic values of Closefast, Closelow, Steady, Openslow and Openfast. The second output is the hot variable which has five triangular membership functions with linguistic values of Closefast, Closelow, Steady, Openslow and Openfast. The diagram of the operation of this controller is illustrated in Fig. 4.10.

This controller is of Mamdani type and uses nine fuzzy rules, which are presented in detailed fashion in Table 4.31.

Fig. 4.10 Diagram of the temperature controller

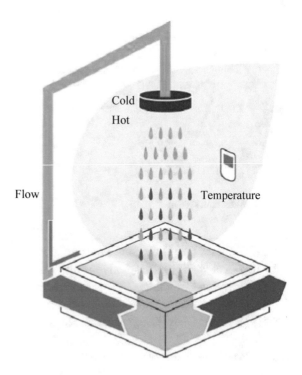

Table 4.31 Fuzzy rules for the temperature controller

Rule number	Inputs			Output	
	Temp	Operator	Flow	Cold	Hot
1	Cold	And	Soft	Openslow	Openfast
2	Cold	And	Good	Closeslow	Openslow
3	Cold	And	Hard	Closefast	Closeslow
4	Good	And	Soft	Openslow	Openslow
5	Good	And	Good	Steady	Steady
6	Good	And	Hard	Closeslow	Closeslow
7	Hot	And	Soft	Openfast	Openslow
8	Hot	And	Good	Openslow	Closeslow
9	Hot	And	Hard	Closeslow	Closefast

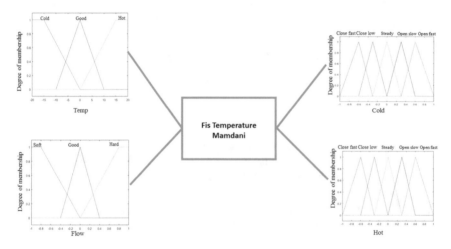

Fig. 4.11 Structure of the temperature fuzzy system for control

Table 4.31 presents in detail fashion each of the rules that compose the temperature controller to achieve the objective of this controller. Figure 4.11 shows the structure of the fuzzy system for this controller.

4.2.1.3 Robot Mobile Controller

The main goal of this controller is to follow a reference trajectory, and it is based on the model of a unicycle mobile robot, which is composed of two drive wheels located on the same axis and a front free wheel that is used only for stability. The graphical representation of this model can be found in Fig. 4.12.

The operation of the robot model is determined by Eqs. 4.7 and 4.8.

Fig. 4.12 Diagram of the
robot mobile controller

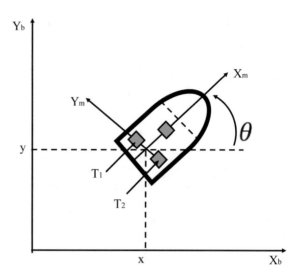

$$M(q)\dot{v} + C(q, \dot{q})v + Dv = \tau + P(t) \tag{4.7}$$

where

$q = (x, y, \theta)^T$ is the vector of the configuration coordinates.
$v = (v,)^T$ is the vector of velocities.
$\tau = (\tau_1, \tau_2)$ is the vector of torques applied to the wheels of the robot where τ_1
                and τ_2 denote the torques of the right and left wheel, respectively.
$P \in R^2$ is the uniformly bounded disturbance vector.
$M(q) \in R^{2X2}$ is the positive-definite inertia matrix.
$C(q, \dot{q})\vartheta$ is the vector of centripetal and Coriolis forces.
$D \in R^{2X2}$ is a diagonal positive-definite damping matrix

The kinematic system is determined by Eq. 4.8:

$$\dot{q} = \begin{bmatrix} \cos \theta & 0 \\ \sin \theta & 0 \\ 0 & 1 \end{bmatrix} \begin{bmatrix} v \\ \omega \end{bmatrix} \tag{4.8}$$

where

(x,y) is the position in the $X - Y$ (world) reference frame
θ is the angle between the heading direction and the x-axis
$v \ and \ \omega$ are the linear and angular velocities, respectively

Equation 4.9 represents the non-holonomic constraint, which in this system corresponds to a non-slip wheel condition preventing the robot from moving sideways.

Table 4.32 Fuzzy rules for the robot mobile controller

Rule number	Inputs			Output	
	e_v	Operator	e_w	T1	T2
1	N	And	N	N	N
2	N	And	Z	N	Z
3	N	And	P	N	P
4	Z	And	N	Z	N
5	Z	And	Z	Z	Z
6	Z	And	P	Z	P
7	P	And	N	P	N
8	P	And	Z	P	Z
9	P	And	P	P	P

$$\dot{y}\cos\theta - \dot{x}\sin\theta = 0 \qquad (4.9)$$

This controller is composed of two inputs and two outputs. The first input is the e_v (error in the linear velocity) which has two trapezoidal and one triangular membership functions with linguistic variables called N, Z and P. The second input is the e_w (error in the angular velocity) which has two trapezoidal and one triangular membership functions with linguistic variables of N, Z and P. The first output is T1 (torque 1), which has the three triangular membership functions with linguistic variables of N, Z and P. The second output is T2 (torque 2), which has the three triangular membership functions with linguistic variables of N, Z and P. This controller is of Mamdani type and uses nine fuzzy rules, which are presented in detail in Table 4.32.

Table 4.32 presents in detail each of the rules that composes the robot mobile controller to achieve the main objective of this controller. Figure 4.13 shows the structure of the fuzzy system for this controller.

4.2.1.4 D.C. Motor Speed Controller

The speed control in a D.C. motor is a classical benchmark control problem with applications in industry. This problem is proposed to be used in this book as a case study because the optimization in a real problem was also of interest, so this experiment aims to demonstrate the capability of the proposed approach to improve the optimization of the controller. In this case, the experiment is focused on the time reduction of the optimization. Figure 4.14 shows the graphical representation of the speed control in a D.C. motor.

The fuzzy system for the speed control in a D.C. motor contains two inputs and one output; input 1 is the error and input 2 is change of the error, and the output is the voltage. The fuzzy system is of Mamdani type, as shown in Fig. 4.15. The fuzzy system contains 15 rules which are shown in Table 4.33.

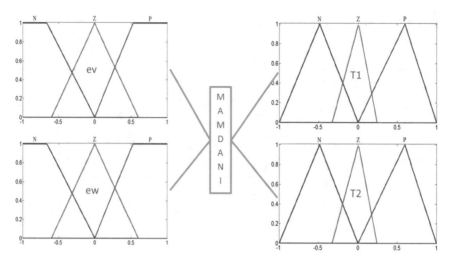

Fig. 4.13 Structure of the mobile robot fuzzy system for control

Fig. 4.14 Graphical representation

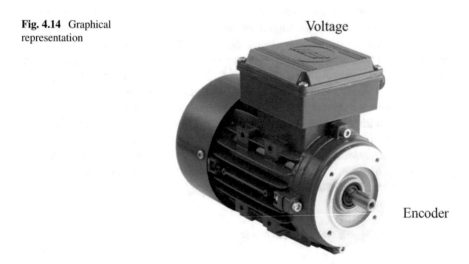

4.2.1.5   Experiments and Results

This section presents the results obtained by optimizing membership functions using the dynamic parameter adaptation with type-1 and interval type-2 fuzzy logic. The methodologies type-1 and interval type-2 fuzzy logic were applied to the water tank, temperature and robot mobile benchmark problems, and for the case of motor speed controller the type-1 methodology was used, as shown in Sect. 4.1.1.1, where the objective is to follow a desired trajectory, and 30 experiments with disturbances and without disturbances were carried out with the original algorithm of harmony search (HS), type-1 fuzzy harmony search algorithm (T1-FHS) and interval type-2 fuzzy

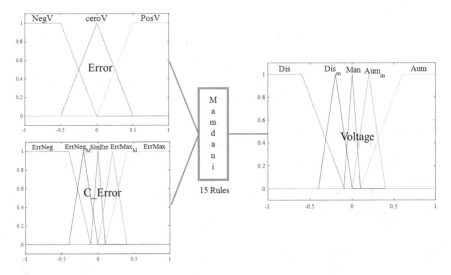

Fig. 4.15 Fuzzy system for motor control

Table 4.33 Fuzzy rules for motor control

No.	Inputs		Output
	Error	Change in Error	Voltage
1	NegV	ErrNeg	Dis
2	NegV	SinErr	Dis
3	NegV	ErrMax	Dis_m
4	CeroV	ErrNeg	Aum_m
5	CeroV	ErrMax	Dis_m
6	PosV	ErrNeg	Aum_m
7	PosV	SinErr	Aum
8	PosV	ErrMax	Aum
9	CeroV	SinErr	Man
10	NegV	ErrNeg_M	Dis
11	CeroV	ErrNeg_M	Aum_m
12	PosV	ErrNeg_M	Aum
13	PosV	ErrMax_M	Aum
14	CeroV	ErrMax_M	Dis_m
15	NegV	ErrMax_M	Dis

harmony search (IT2-FHS). The noise applied to each plant was with a level of 0.05 using the uniform random number generator. The objective function of the methods is the root mean square error (RMSE) of the trajectory, and is given by Eq. 4.10.

$$RMSE = \sqrt{\frac{1}{N} \sum_{t=1}^{N} (x_t - \hat{x}_t)^2} \qquad (4.10)$$

The parameters used to perform the experiments are as follows: 30 iterations, PArate 0.75, *HMR* 0.95 (only for the HS) and HM50.

The experimental results in Tables 4.34, 4.35, 4.36, and 4.37 present the best experiment without perturbation in the fuzzy logic controller for each algorithm applied to each case study (water tank controller, temperature controller, robot mobile controller and motor speed controller). Tables 4.38, 4.39, 4.40 and 4.41 show the results obtained using the water tank controller, temperature controller, robot mobile controller and motor speed controller with noise. These results are of the average of the 30 experiments for each method.

Table 4.34 Errors obtained from the water tank controller without noise

Performance index	HS		T1-FHS		IT2-FHS	
	Mean	S.D.	Mean	S.D.	Mean	S.D.
RMSE (Mean)	3.18E–02	3.13E–02	2.61E–01	1.55e–02	2.49E–02	1.97e–02

Table 4.35 Errors obtained from the temperature controller without noise

Performance index	HS		T1-FHS		IT2-FHS	
	Mean	S.D.	Mean	S.D.	Mean	S.D.
RMSE (MEAN)	6.34E–02	3.12E–03	6.75E–02	1.59E–03	6.25E–02	1.13E–03

Table 4.36 Errors obtained from the robot mobile controller without noise

Performance index	HS		T1-FHS		IT2-FHS	
	Mean	S.D.	Mean	S.D.	Mean	S.D.
RMSE (Mean)	2.33E–01	2.18E–01	1.37E–01	1.57E–01	1.11E–01	1.37E–01

Table 4.37 Errors obtained from the motor speed controller without noise

Performance index	HS		T1-FHS	
	Mean	S.D.	Mean	S.D.
RMSE (Mean)	5.21E–01	3.33E–02	5.78E–01	7.77E–02

Table 4.38 Errors obtained from the water tank controller with noise

Performance index	HS		T1-FHS		IT2-FHS	
	Mean	S.D.	Mean	S.D.	Mean	S.D.
RMSE (Mean)	3.18E–02	3.13E–02	2.49E–02	2.01E–02	1.32E–02	1.54E–02

Table 4.39 Errors obtained from the temperature mobile controller with noise

Performance index	HS		T1-FHS		IT2-FHS	
	Mean	S.D.	Mean	S.D.	Mean	S.D.
RMSE (Mean)	2.33E–01	3.12E–03	3.82E–02	3.34E–03	3.69E–02	1.07E–04

Table 4.40 Errors obtained from the robot mobile controller with noise

Performance index	HS		T1-FHS		IT2-FHS	
	Mean	S.D.	Mean	S.D.	Mean	S.D.
RMSE (Mean)	6.34E–02	2.18E–01	1.24E–01	4.22E–02	6.29E–04	4.20E–02

Table 4.41 Errors obtained from the motor speed controller with noise

Performance index	HS		T1-FHS	
	Mean	S.D.	Mean	S.D.
RMSE (Mean)	5.35E–01	2.95E–02	4.20E–05	8.43E–05

Analyzing the results obtained from Tables 4.34, 4.35, 4.36, 4.37, 4.38, 4.39, 4.40 and 4.41, we can identify an improvement in the experiments when using the type-1 and type-2 methods with noise. Figures 4.16 and 4.17 show the best-found desired trajectories with each method for the water tank controller.

Figures 4.18 and 4.19 show the best-found desired trajectories with each method for the temperature controller.

Figures 4.20 and 4.21 show the best-found desired trajectories with each method for the robot mobile controller.

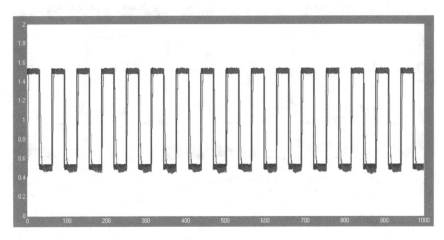

Fig. 4.16 The best result using the T1-FHS with noise

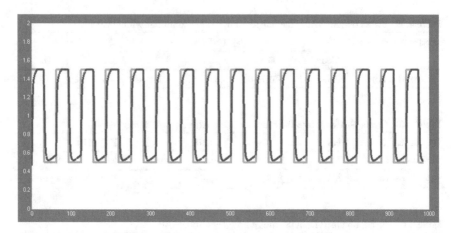

Fig. 4.17   The best result using the IT2-FHS with noise

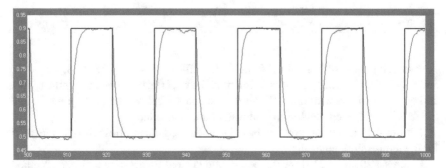

Fig. 4.18   The best result using the T1-FHS with noise

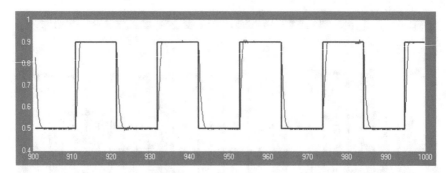

Fig. 4.19   The best result using the IT2-FHS with noise

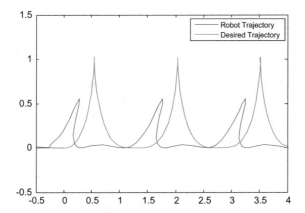

Fig. 4.20 The best result using the T1-FHS with noise

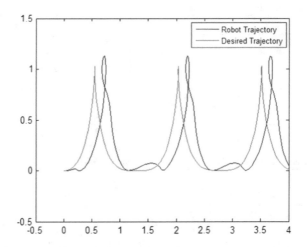

Fig. 4.21 The best result using the IT2-FHS with noise

Figure 4.22 shows the best-found desired trajectory with T1-FHS method for the motor speed controller.

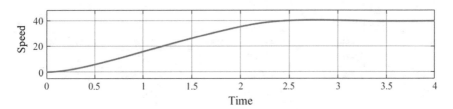

Fig. 4.22 The best result using the T1-FHS with noise

4.2.1.6 Statistical Test

This section shows a statistical comparison between the original method and the proposed methods applied to the four control problems mentioned above. The statistical test that was applied is the z-test, which is given by Eq. 4.4, and the parameters used are shown in Table 4.42.

The alternative hypothesis indicates that the type-1 FHS (T1-FHS) and type-2 FHS (IT2-FHS) methods are smaller than the original harmony search (HS) algorithm method and the null hypothesis indicates otherwise, with rejection region for values less than −1.645.

Tables 4.43, 4.44 and 4.45 show the values of Z, "*S*" means that evidence of significance is found and "*N.S.*" refers to no evidence of significance is found. The result in the first row represents the comparison between FHS and HS; the result in

Table 4.42 Parameters for the statistical z-test

Parameter	Value
Level of confidence	95%
Alpha	0.05%
H_a	$\mu_1 < \mu_2$
H_0	$\mu_1 \geq \mu_2$
Critical value	−1.645

Table 4.43 Results for the statistical test with type-1 and type-2 FHS and HS without noise

Water tank controller				
Method	Mean	Standard deviation	Z-value	Evidence
HS	3.19E–02	3.13E–02	35.8789	*N.S.*
FHS	2.60E–01	1.55e–02	−0.9330	*N.S.*
FHS2	2.56E–02	1.97e–02	−51.21	S
Temperature controller				
Method	Mean	Standard deviation	Z-value	Evidence
HS	6.34E–02	3.12E–03	6.4129	*N.S.*
FHS	6.75E–02	1.59E–03	−1.4855	*N.S.*
FHS2	6.25E–02	1.13E–03	−14.0396	S
Robot mobile controller				
Method	Mean	Standard deviation	Z-value	Evidence
HS	2.33E–01	2.18E–01	−1.9572	S
FHS	1.37E–01	1.57E–01	−2.5953	S
FHS2	1.11E–01	1.37E–01	−0.6834	*N.S.*

Table 4.44 Results for the statistical test with type-1 and type-2 FHS and HS with noise

Water tank controller

Method	Mean	Standard deviation	Z-value	Evidence
HS	3.18E–02	3.13E–02	−1.0160	N.S.
FHS	2.49E–02	2.01E–02	−2.9205	S
FHS2	1.32E–02	1.54E–02	−2.5308	S

Temperature controller

Method	Mean	Standard deviation	Z-value	Evidence
HS	6.34E–02	3.12E–03	72.6213	N.S.
FHS	1.24E–01	3.34E–03	−110.1311	S
FHS2	6.29E–04	1.07E–04	−202.2109	S

Robot Controller

Method	Mean	Standard deviation	Z-value	Evidence
HS	2.33E–01	2.18E–01	−4.8051	S
FHS	3.82E–02	4.22E–02	−4.8380	S
FHS2	3.69E–02	4.20E–02	−0.1196	N.S.

Table 4.45 Results for the statistical test with type-1 and HS

D.C. Motor speed controller

Method	Mean	Standard deviation	Z-value	Evidence
HS without noise	5.21E–01	3.33E–02	1.7237	N.S.
HS with noise	5.35E–01	2.95E–02		
FHS without noise	5.78E–01	7.77E–02	−40.7414	S
FHS with noise	4.20E–05	8.43E–05		

the second row represents the comparison between FHS2 and HS, and finally, the result in the third row is the comparison between FHS and FHS2.

4.2.2 Sugeno Controller

The Takagi-Sugeno-Kang (TSK) model which is an alternative model in which the consequent does not give us a fuzzy set but a linear function. The main difference between the TSK method and the Mamdani method is that in TSK it is not necessary to perform a defuzzifier process.

4.2.2.1   Ball and Beam Controller

The objective of this control model is to achieve the balance of the ball in the beam, where it is allowed to roll with one degree of freedom along the length of the beam. A lever arm is attached to the beam at one end and a servo gear at the other. As the servo gear turns by an angle θ, the lever changes the angle of the beam by α. When the angle is changed from the horizontal position, gravity causes the ball to roll along the beam. According to the Euler–Lagrange method, the mathematical model of motion for the ball and beam system is presented in Eq. 4.11 and the beam angle (α) can be expressed in terms of the angle of the gear (θ), presented in Eq. 4.12.

$$0 = \left(\frac{J}{R^2} + m \right) \ddot{r} + mg \sin \alpha - mr \dot{\alpha}^2 \qquad (4.11)$$

$$\alpha = \frac{d}{L} \theta \qquad (4.12)$$

A controller is designed for this system so that the ball's position can be manipulated. The representation of this is illustrated in Fig. 4.23.

The variables and constants in this case are defined in Table 4.46.

This controller contains four inputs, which are: r, $d/dt\ r$, and d/dt , respectively, and are granulated into two generalized bell form membership functions which are small and large. The outputs are linear functions composed of 16 linear equations. Figure 4.24 shows the structure of the fuzzy system for this controller. This controller is of Takagi-Sugeno type and uses 16 fuzzy rules, which are presented with more detail in Fig. 4.25.

The output of the ball and beam controller contains 16 linear functions that are described in Fig. 4.26.

Fig. 4.23 Process of the ball and beam (BB) system

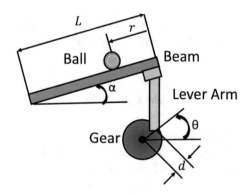

Table 4.46 Variables of ball and beam system

Symbol	Definition	Value
m	Mass of the ball	0.11 kg
R	Radius of the ball	0.015 m
d	Lever arm offset	0.03 m
g	Gravitational acceleration	9.8 m/s^2
L	Length of the beam	1.0 m
J	Ball's moment of inertia	9.9 e–6 kg m^2
r	Ball position coordinate	
α(alpha)	Beam angle coordinate	
θ (Theta)	Servo gear angle	

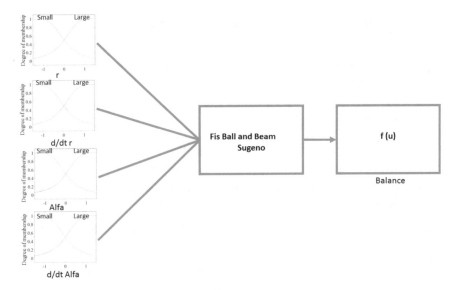

Fig. 4.24 Graphical representation of the ball and beam fuzzy system controller

4.2.2.2 Inverted Pendulum Controller

The inverted pendulum on a cart problem is a complex nonlinear control problem, with its complexity originating from the nonlinear nature of the plant. In the cart-pole system the pendulum rod is free to oscillate around a fixed pivot point attached to a cart, which is controlled by a motor and is constrained to move in the horizontal direction. When the rod is placed in the upright vertical position, it is in an unstable equilibrium point. The main goal of the controller is to apply a force to move the cart so that the pendulum remains in the vertical unstable position. The cart-pole system is shown in Fig. 4.27, where x is the cart position, θ (theta) is the pendulum angle and F is the control force applied, parallel to the rail, to the cart, where g = 9.81 m/s^2.

1. If r is Small and d/dt r is Small and alfa is Small and d/dt alfa is Small then Out = out1

2. If r is Small and d/dt r is Small and alfa is Small and d/dt alfa is Large then Out = out2

3. If r is Small and d/dt r is Small and alfa is Large and d/dt alfa is Small then Out = out3

4. If r is Small and d/dt r is Small and alfa is Large and d/dt alfa is Large then Out = out4

5. If r is Small and d/dt r is Large and alfa is Small and d/dt alfa is Small then Out = out5

6. If r is Small and d/dt r is Large and alfa is Small and d/dt alfa is Large then Out = out6

7. If r is Small and d/dt r is Large and alfa is Large and d/dt alfa is Small then Out = out7

8. If r is Small and d/dt r is Large and alfa is Large and d/dt alfa is Large then Out = out8

9. If r is Large and d/dt r is Small and alfa is Small and d/dt alfa is Small then Out = out9

10. If r is Large and d/dt r is Small and alfa is Small and d/dt alfa is Large then Out = out10

11. If r is Large and d/dt r is Small and alfa is Large and d/dt alfa is Small then Out = out11

12. If r is Large and d/dt r is Small and alfa is Large and d/dt alfa is Large then Out = out12

13. If r is Large and d/dt r is Large and alfa is Small and d/dt alfa is Small then Out = out13

14. If r is Large and d/dt r is Large and alfa is Small and d/dt alfa is Large then Out = out14

15. If r is Large and d/dt r is Large and alfa is Large and d/dt alfa is Small then Out = out15

16. If r is Large and d/dt r is Large and alfa is Large and d/dt alfa is Large then Out = out16

Fig. 4.25 Fuzzy rules for the ball and beam controller

out(1)=1.015r+ 2.234 r^'- 12.67α - 4.046α^' + 0.02624

out(2)=1.161r + 1.969r^' - 9.396α - 6.165α^' + 0.474

out(3)= 1.506r + 2.234r^'- 12.99α - 1.865α^' + 1.426

out(4)= 0.7339r + 1.969r^' - 9.381α - 4.688α^' - 0.8804

out(5)= 0.7343r + 2.234 r^' - 12.85α - 6.11α^' - 1.034

out(6)= 1.413r + 1.969 r^' - 9.485α - 6.592α^' + 1.159

out(7)= 1.225r + 2.234 r^' - 12.8α - 3.929α^' + 0.3662

out(8)= 0.9853r + 1.969r^' - 9.291α - 5.115α^' - 0.195

out(9)= 0.9853r + 1.969 r^' - 9.292α – 5.115α^'+ 0.195

out(10)= 1.225r + 2.234 r^' - 12.8α – 3.929α^' - 0.3662

out(11)= 1.413r + 1.969 r^' - 9.485α - 6.592α^' - 1.159

out(12)= 0.7343r + 2.234 r^' - 12.85α - 6.11α^' + 1.034

out(13)= 0.7339r + 1.969r^' - 9.381α - 4.688α^' + 0.8804

out(14)= 1.506r + 2.234 r^' - 12.99α - 1.865α^' - 1.426

out(15)= 1.161r + 1.969r^' - 9.396α – 6.165α^' - 0.474

out(16)= 1.015r + 2.234 r^' - 12.67α - 4.046α^' - 0.02624

Fig. 4.26 Sugeno coefficients in the outputs

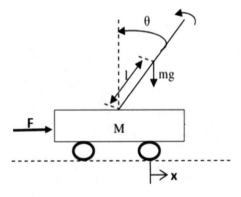

Fig. 4.27 Inverted pendulum system

The fuzzy system for this problem is of Takagi-Sugeno type. It used the max aggregation and the wtaver defuzzification, which is composed of four inputs with triangular membership functions and one output with 16 linear functions, as represented in Fig. 4.28, and the fuzzy rules that are considered are shown in Fig. 4.29.

The fuzzy rules that are illustrated in Fig. 4.29 are those that maintain control of the cart for the inverted pendulum. The combination of the rules of Fig. 4.29 is given by the following meanings: in1 = Pendulum angle, in2 = Angular velocity, in3 = Car position, in4 = Car velocity, in1mf1 = Small, in1mf2 = Large, in2mf1 = Small, in2mf2 = Large, in3mf1 = Small, in3mf2 = Large, in4mf1 = Small, in4mf2 = Large.

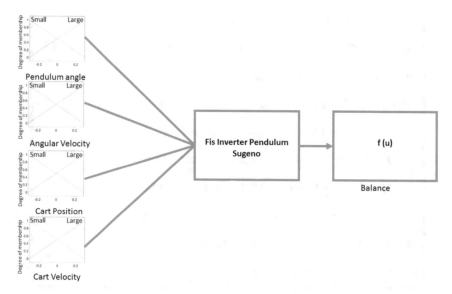

Fig. 4.28   Structure of the fuzzy system for the inverted pendulum

```
1. If (in1 is in1mf1) and (in2 is in2mf1) and (in3 is in3mf1) and (in4 is in4mf1) then (out is outmf1) (1)
2. If (in1 is in1mf1) and (in2 is in2mf1) and (in3 is in3mf1) and (in4 is in4mf2) then (out is outmf2) (1)
3. If (in1 is in1mf1) and (in2 is in2mf1) and (in3 is in3mf2) and (in4 is in4mf1) then (out is outmf3) (1)
4. If (in1 is in1mf1) and (in2 is in2mf1) and (in3 is in3mf2) and (in4 is in4mf2) then (out is outmf4) (1)
5. If (in1 is in1mf1) and (in2 is in2mf2) and (in3 is in3mf1) and (in4 is in4mf1) then (out is outmf5) (1)
6. If (in1 is in1mf1) and (in2 is in2mf2) and (in3 is in3mf1) and (in4 is in4mf2) then (out is outmf6) (1)
7. If (in1 is in1mf1) and (in2 is in2mf2) and (in3 is in3mf2) and (in4 is in4mf1) then (out is outmf7) (1)
8. If (in1 is in1mf1) and (in2 is in2mf2) and (in3 is in3mf2) and (in4 is in4mf2) then (out is outmf8) (1)
9. If (in1 is in1mf2) and (in2 is in2mf1) and (in3 is in3mf1) and (in4 is in4mf1) then (out is outmf9) (1)
10. If (in1 is in1mf2) and (in2 is in2mf1) and (in3 is in3mf1) and (in4 is in4mf2) then (out is outmf10) (1)
11. If (in1 is in1mf2) and (in2 is in2mf1) and (in3 is in3mf2) and (in4 is in4mf1) then (out is outmf11) (1)
12. If (in1 is in1mf2) and (in2 is in2mf1) and (in3 is in3mf2) and (in4 is in4mf2) then (out is outmf12) (1)
13. If (in1 is in1mf2) and (in2 is in2mf2) and (in3 is in3mf1) and (in4 is in4mf1) then (out is outmf13) (1)
14. If (in1 is in1mf2) and (in2 is in2mf2) and (in3 is in3mf1) and (in4 is in4mf2) then (out is outmf14) (1)
15. If (in1 is in1mf2) and (in2 is in2mf2) and (in3 is in3mf2) and (in4 is in4mf1) then (out is outmf15) (1)
16. If (in1 is in1mf2) and (in2 is in2mf2) and (in3 is in3mf2) and (in4 is in4mf2) then (out is outmf16) (1)
```

Fig. 4.29   Fuzzy rules for the inverted pendulum

$outmf_1$: 41.37 **in1** + 10.03 **in2** + 3.162 **in3** + 4.288 **in4** + 0.3386

$outmf_2$: 40.41 **in1** + 10.05 **in2** + 3.162 **in3** + 4.288 **in4** + 0.2068

$outmf_3$: 41.37 **in1** + 10.03 **in2** + 3.162 **in3** + 4.288 **in4** + 0.3386

$outmf_4$: 41.41 **in1** + 10.05 **in2** + 3.162 **in3** + 4.288 **in4** + 0.2068

$outmf_5$: 38.56 **in1** + 10.18 **in2** + 3.162 **in3** + 4.288 **in4** − 0.04893

$outmf_6$: 37.6 **in1** + 10.15 **in2** + 3.162 **in3** + 4.288 **in4** − 0.1807

$outmf_7$: 38.56 **in1** + 10.18 **in2** + 3.162 **in3** + 4.288 **in4** − 0.04893

$outmf_8$: 37.6 **in1** + 10.15 **in2** + 3.162 **in3** + 4.288 **in4** − 0.1807

$outmf_9$: 37.6 **in1** + 10.15 **in2** + 3.162 **in3** + 4.288 **in4** + 0.1807

$outmf_{10}$: 38.56 **in1** + 10.18 **in2** + 3.162 **in3** + 4.288 **in4** + 0.04891

$outmf_{11}$: 37.6 **in1** + 10.15 **in2** + 3.162 **in3** + 4.288 **in4** + 0.1807

$outmf_{12}$: 38.56 **in1** + 10.18 **in2** + 3.162 **in3** + 4.288 **in4** + 0.04892

$outmf_{13}$: 40.41 **in1** + 10.05 **in2** + 3.162 **in3** + 4.288 **in4** − 0.2068

$outmf_{14}$: 41.37 **in1** + 10.03 **in2** + 3.162 **in3** + 4.288 **in4** − 0.3386

$outmf_{15}$: 40.41 **in1** + 10.05 **in2** + 3.162 **in3** + 4.288 **in4** − 0.2068

$outmf_{16}$: 41.37 **in1** + 10.03 **in2** + 3.162 **in3** + 4.288 **in4** − 0.3386

Fig. 4.30 Sugeno coefficients in the outputs

They are representing the 16 Sugeno coefficients in the outputs corresponding to the combination of the rules shown in Fig. 4.30:

Where in1 represents the pendulum angle, in2 represents the angular velocity, in3 represents the cart position and in4 represents the cart velocity.

4.2.2.3 Experiments and Results

The results obtained by optimizing the input membership functions using the proposed method with T1-FHS, IT2-FHS and GT2-FHS fuzzy system are presented. This methodology was applied to the ball and beam controller, where the goal is to maintain the balance of the ball in the beam. For the inverted pendulum controller only the T1-FHS is used. It is important to emphasize that the exits of the membership functions for the two controllers are fixed throughout the iterations.

A total of 30 experiments with the HS method have been performed and also experiments with perturbation and without perturbation in the controller were carried out with the FHS, IT2-FHS and GT2-FHS fuzzy harmony search algorithms.

The noise applied for this controller is Gaussian distribution with a level of 0.05. In this case it is decided to use this type of noise since in the previous work we obtained favorable results by applying this same level of perturbation to the water tank, temperature and the mobile robot controller. The root mean square error (RMSE) is the objective function of the boot controllers, although there are other control metrics that were also used which are the following: ISE (integral of squared error), IAE (integral of the absolute value of the error), ITSE (integral of time-weighted squared error), ITAE (integral of the time multiplied by the absolute value of the error) and SNR (signal-to-noise ratio), respectively, presented in Eqs. 4.13–4.17.

$$ISE = \sum_{t=1}^{N} \left| (x(t) - \hat{x}(t))^2 \right| \qquad (4.13)$$

$$IAE = \sum_{t=1}^{N} \left| x(t) - \hat{x}(t) \right| \qquad (4.14)$$

$$ITSE = \sum_{t=1}^{N} t \left((x(t) - \hat{x}(t))^2 \right) \qquad (4.15)$$

$$ITAE = \sum_{t=1}^{N} t \left(x(t) - \hat{x}(t) \right) \qquad (4.16)$$

$$SNR_{dB} = 10\log_{10} \left(\frac{P_{signal}}{P_{noise}} \right) \qquad (4.17)$$

The parameters used are as follows: 100 iterations, *PArate* 0.75, *HMR* 0.95 (only for the HS, in the T1-FHS, IT2-FHS and GT2-FHS methods this parameter changes dynamically with fuzzy system), and HM 30. Table 4.47 shows the simulation results obtained using the ball and beam controller without perturbation in the controller and Table 4.48 shows the simulation results obtained using the ball and beam controller under perturbations.

Table 4.47 Simulation results without perturbation in the controller

Performance index	Methods			
	HS	T1-FHS	IT2-FHS	GT2-FHS
RMSE (Mean)	1.53E+00	1.57E+00	1.56E+00	**1.52E+00**
S.D.	2.64E–02	1.33E–02	1.46E–02	1.99E–02
ITAE	8.87E+05	1.01E+06	1.01E+06	**8.39E+05**
ITSE	2.02E+06	2.62E+06	2.62E+06	**1.67E+06**
IAE	1.78E+03	2.02E+03	2.02E+03	**1.68E+03**
ISE	4.06E+03	5.25E+03	5.25E+03	**3.34E+03**

Table 4.48 Results with perturbation in the controller

Performance index	Methods		
	T1-FHS	IT2-FHS	GT2-FHS
RMSE (Mean)	1.55E+00	1.52E+00	**1.49E+00**
S.D.	1.31E–02	3.21E–02	8.22E–03
ITAE	1.00E+06	1.01E+06	**8.09E+05**
ITSE	2.59E+06	2.63E+06	**1.50E+06**
IAE	2.01E+03	2.03E+03	**1.62E+03**
ISE	5.18E+03	5.26E+03	**3.01E+03**

Table 4.47 summarizes the results obtained for each method without perturbation: fuzzy HS with type-1 (T1-FHS), fuzzy HS with interval type-2 (IT2-FHS), fuzzy HS with generalized type-2 (GT2-FHS).

Table 4.48 shows the results obtained with 30 carried out experiments for each method with perturbation: fuzzy HS with type-1 (T1-FHS), fuzzy HS with interval type-2 (IT2-FHS) and fuzzy HS with generalized type-2 (GT2-FHS).

Figure 4.31 illustrates the behavior of equilibrium for each method without noise in the controller, and Fig. 4.32 shows the behavior of equilibrium for each method

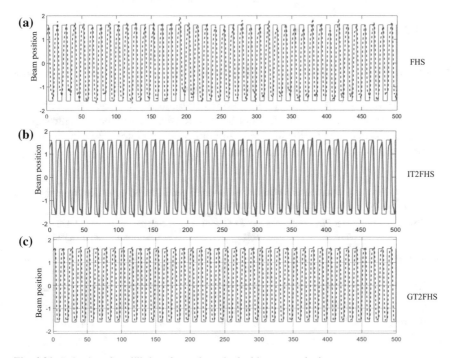

Fig. 4.31 Behavior of equilibrium for each method without perturbation

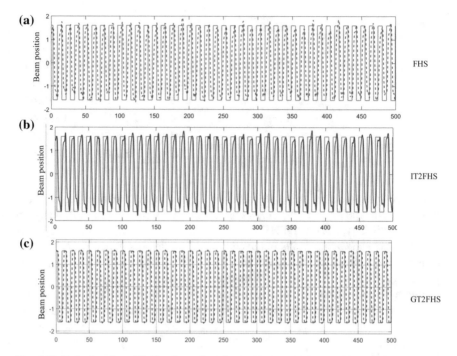

Fig. 4.32 Behavior of equilibrium for each method with perturbation

with noise in the controller: (a) T1-FHS, (b) IT2-FHS and (c) GT2-FHS. The blue line represents the set point and the pink line represents the ball position.

It can be observed that when applying noise to the controller the generalized type-2 FHS method achieved to obtain a better balance for the ball and beam controller.

The results obtained for the inverted pendulum controller are shown in Table 4.49 where the is to minimize the RMSE error.

Table 4.49 Summary of the results obtained for inverted pendulum on a cart controller using HS and T1-FHS algorithms without noise and with noise

RMSE	HS FLC without noise	HS FLC with noise	T1-FHS FLC without noise	T1-FHS FLC with noise
Best	3.15E–01	3.97E–01	2.99E–01	2.97E–01
Worst	4.88E+00	1.28E+00	2.01E+00	1.76E+00
Average	1.88E+00	1.02E+00	7.67E–01	7.54E–01
Standard deviation	1.35E+00	4.07E–01	4.81E–01	4.49E–01

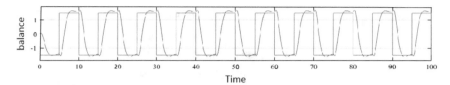

Fig. 4.33 Behavior of balance without perturbation

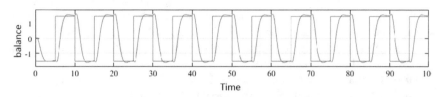

Fig. 4.34 Behavior of balance with perturbation

Table 4.49 shows the results obtained from the inverted pendulum controller by HS and T1-FHS algorithms. In this case only the best, worst, mean and standard deviation of the obtained experiments are shown.

Finally, from the results obtained for the HS algorithm, in inverted pendulum controller, we can see that there is an improvement compared to the original algorithm.

Figures 4.33 and 4.34 illustrated the behavior of balance for T1-FHS without noise and T1-FHS with noise, respectively, in the controller. The blue line represents the set point and the orange line represents the balance.

4.2.2.4 Statistical Test

To validate the effectiveness of the proposed method, a statistical test was applied; in this case the z-test was used.

The claim, parameters and data for the hypothesis testing are as follows:

$$\mu_1 = Proposed\ method\ with\ \text{perturbation}$$
$$\mu_2 = Other\ method\ without\ \text{perturbation}$$

Sample 1 (μ_1) would be the GT2-FHS, IT2-FHS and T1-FHS methods with perturbation and sample 2 (μ_2) would be the GT2-FHS, IT2-FHS and T1-FHS methods without perturbation. The mean of the proposed method with perturbation is less than the mean of the other method without perturbation (Claim). The hypotheses would be as follows:

$$H_o\ :\ \mu_1 \geq \mu_2$$

$$H_a : \mu_1 < \mu_2 (Claim)$$
$$\alpha = 0.05$$
$$\text{Level of confidence} = 95\%$$
$$\text{Sample Size (n)} = 30$$
$$Critical\ Value(Z_o) = -1.645$$

Once the Z-value has been calculated, we can say that there is enough evidence for all values below critical value Zo or −1.645 we can reject the null hypothesis and we can state that the mean of the proposed method with perturbation is lower than the mean of the other method without perturbation with a 95% of confidence, and this Z-value is shown in Table 4.50.

Tables 4.50 and 4.51 show the values of Z, and in these tables "S" means evidence of significance is found and "N.S." refers to which no significant evidence is found.

Table 4.50 shows the results of the mean, standard deviation and the Z-value obtained. In three cases significant evidence is obtained when comparing the proposed method with perturbation against the method without perturbation. With this

Table 4.50 Results for the statistical test of the ball and beam controller

T1-FHS with perturbation		T1-FHS without perturbation		Z-value	Evidence
Mean	S.D.	Mean	S.D.	−5.8680	S
1.55E+00	1.31E–02	1.57E+00	1.33E–02		
IT2-FHS with perturbation		IT2-FHS without perturbation			
Mean	S.D.	Mean	S.D.	−6.2128	S
1.52E+00	3.21E–02	1.56E+00	1.46E–02		
GT2-FHS with perturbation		GT2-FHS without perturbation			
Mean	S.D.	Mean	S.D.	−10.1756	S
1.49E+00	8.22E–03	1.53E+00	1.99E–02		

Table 4.51 Results for the statistical test of the inverted pendulum controller

Controller	μ_1				μ_2	Z-value	Evidence
Inverted pendulum	T1-FHS FLC without noise				HS without noise	−4.25	S
	Mean	S.D.	Mean	S.D.			
	7.67E–01	4.81E–01	1.88E+00	1.35E+00			
	T1-FHS FLC with noise				HS with noise	−2.40	S
	Mean	S.D.	Mean	S.D.			
	7.54E–01	4.49E–01	1.02E+00	4.07E–01			

Table 4.52 Comparison results for the inverted pendulum controller

	FBCO FLC without noise	FBCO FLC with noise	FDE FLC Without noise	FDE FLC with noise	FHS FLC without noise	FHS FLC with noise	Z-value	Evidence
Average	8.21E–02	4.82E–01	2.15E–01	3.78E–01	7.67E–01	7.54E–01	4.2241	N.S.
S.D.	4.66E–01	9.19E–02	2.75E–01	1.90E–01	4.81E–01	4.49E–01	−36.6416	S

we can verify that when using uncertainty, a better stability is achieved in the ball and beam controller.

Table 4.51 shows the results of the evidence of the first row that represents the comparison between the proposed type-1 and the original methods without noise and the result of the evidence on the second row represents the comparison between the proposed type-1 and the original methods with noise.

Table 4.51 shows the results of the Z-value obtained. In two cases significant evidence is obtained when comparing the proposed method with perturbation against the method without perturbation. With this we can verify that when using uncertainty, a better stability is achieved in the inverted pendulum controller.

To validate the effectiveness of the method, it is decided to compare the results with two methods that exist in the literature [10], which are the fuzzy differential evolution (FDE) and fuzzy bee colony optimization (FBCO) where the inverted pendulum plant is used and the results are shown in Table 4.52.

Tables 4.52 shows the values of Z, "S" means that evidence of significance is found and "N.S." refers to which no evidence is found of significance. The result in the first row represents the comparison between FHS with noise and FDE with noise finally the result with the second row represents the comparison between FHS with noise and FBCO with noise.

Table 4.52 shows the mean and standard deviation of the 30 experiments obtained with the FBCO and FDE methods and the statistical test where we can see that the order of the results with the methods is the same as those obtained by the proposed method, which indicates that the algorithm obtains a good balance for this driver. When comparing the means of the three methods we can see that the FDE method obtains a better RMSE as shown in the following figures. The results of the statistical test indicate that comparing the proposed method against the FDE does not achieve significant evidence, but comparing the results with the FBCO significant evidence is obtained.

Figure 4.35 shows that DE found better results compared to BCO and HS with the original and fuzzy algorithms for the inverted pendulum on a cart controller.

With the goal to analyze the behavior of each algorithm, Fig. 4.36 shows a comparison of each fuzzy algorithm with noise in the model for the inverted pendulum on a cart controller.

Figure 4.36 shows that DE algorithm is better when noise is added in the model compared to BCO and HS in both benchmark problems.

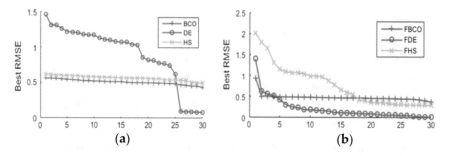

Fig. 4.35 Comparison of the RMSE for each method for the inverted pendulum on a cart controller: **a** Original algorithms, **b** Fuzzy algorithm

Fig. 4.36 Comparison of the RMSE for each method with noise

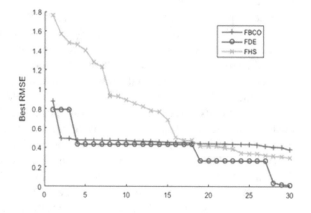

References

1. Peraza, C., Valdez, F., Garcia, M., Melin, P., Castillo, O.: A new fuzzy harmony search algorithm using fuzzy logic for dynamic parameter adaptation. Algorithms **9**(4), 69 (2016)
2. Peraza, C., Valdez, F., Castillo, O.: Interval type-2 fuzzy logic for dynamic parameter adaptation in the Harmony search algorithm, pp. 106–112. (2016)
3. Peraza, C., Valdez, F., Castillo, O.: Fuzzy harmony search algorithm using an interval type-2 fuzzy logic applied to benchmark mathematical functions. In: Hadjiski, M., Atanassov, K.T. (eds.) Intuitionistic Fuzziness and Other Intelligent Theories and Their Applications, vol. 757, pp. 13–28. Springer International Publishing, Cham (2019)
4. Wang, C.-M., Huang, Y.-F.: Self-adaptive harmony search algorithm for optimization. Expert Syst. Appl. **37**(4), 2826–2837 (2010)
5. Omran, M.G.H., Mahdavi, M.: Global-best harmony search. Appl. Math. Comput. **198**(2), 643–656 (2008)
6. Cheng, S., Qin, Q., Wu, Z., Shi, Y., Zhang, Q.: Multimodal optimization using particle swarm optimization algorithms: CEC 2015 competition on single objective multi-niche optimization. In: *2015 IEEE Congress on Evolutionary Computation (CEC)*, pp. 1075–1082. Sendai, Japan (2015)
7. Gu, F., Cheung, Y., Luo, J.: An evolutionary algorithm based on decomposition for multimodal optimization problems. In: *2015 IEEE Congress on Evolutionary Computation (CEC)*, pp. 1091–1097. Sendai, Japan (2015)

8. Zheng, S., Yu, C., Li, J., Tan, Y.: Exponentially decreased dimension number strategy based dynamic search fireworks algorithm for solving CEC2015 competition problems. In: *2015 IEEE Congress on Evolutionary Computation (CEC)*pp. 1083–1090. Sendai, Japan, (2015)

9. Caraveo, C., Valdez, F., Castillo, O.: A new optimization meta-heuristic algorithm based on self-defense mechanism of the plants with three reproduction operators. Soft. Comput. **22**(15), 4907–4920 (2018)

10. Castillo, O., Valdez, F., Soria, J., Amador-Angulo, L., Ochoa, P., Peraza, C.: Comparative study in fuzzy controller optimization using bee colony, differential evolution, and harmony search algorithms. Algorithms **12**(1), 9 (2018)

Chapter 5
Conclusions to Fuzzy Harmony Search

In this book we propose a new optimization metaheuristic that is inspired by the original harmony search algorithm. This algorithm was created recently and we have successfully achieved the integration of fuzzy logic for dynamic parameter adaptation using type-1, interval type-2 and generalized type-2 fuzzy systems.

The main difference between the original algorithm and the proposed is that the original algorithm uses fixed parameters along the iterations and the proposed algorithm uses fuzzy logic to adjust the parameters as the iterations progress monitoring by rules exploration and exploitation of the search space in order to find better results when applied to case studies.

The main contribution of this book is to solve this problem applied to the original harmony search algorithm using type-1, interval type-2 and generalized type-2 fuzzy logic. The proposed methodology was applied to the resolution of problems of classical benchmark mathematical functions, CEC 2015, CEC 2017 functions with different dimensions and to the optimization of various fuzzy logic control cases. The parameter adaptation is responsible for optimizing the values of input and output membership functions of the Mamdani control problems and the input membership function of the Sugeno control problems, where the objective is to minimize the root mean square error.

This methodology was applied with the original HS algorithm and with the proposed FHS algorithms with noise and without noise for the controllers. A perturbation of 0.05 was used to check the stability of the methods, since theoretically if more uncertainty is handled in a problem it must have more stability. The results obtained from the 30 experiments for each method show that when perturbation is applied stability is achieved, and we can observe that interval type-2 and generalized type-2 fuzzy systems with perturbation or without perturbation for the controller were used.

In Chap. 2, an exhaustive analysis of the algorithm of the original search was performed, testing all the parameters that affect the operation of the algorithm such as iterations, harmonies, dimensions, HMR and stopping criterion, in order to obtain knowledge of the best parameters to apply them in our proposal and try using different cases.

In Chap. 3, the proposed methodology is presented once the behavior of the algorithm has been validated in Chap. 2, and it was possible to obtain the knowledge to create the inputs, outputs and rules of the fuzzy system that will be in charge of adjusting the parameters as the number of iterations progress. Each parameter, HMR, PArate, was tested separately with different rules in increment, decrement; varying the ranges of the outputs of the fuzzy system; combinations of rules and inputs and outputs were also made.

Chapter 4 shows the proposed fuzzy systems of type-1, type-2 and generalized type 2, consisting of one input and one output with three rules in increment applied to 10 mathematical benchmark classic functions with 10, 30 and 100 dimensions, respectively. Two comparisons were made with existing articles in the literature, and it was found that by using a larger number of dimensions the method achieves good results. Statistically there is significant evidence in most of the functions presented. Also, in this chapter, experiments were carried out with 15 mathematical functions of CEC 2015 benchmark using the original method and the proposed type and interval type-2 methods with 10, 30, 50 and 100 dimensions, obtaining evidence in some functions compared to the original method. Also, in this chapter, experiments were carried out with 30 mathematical functions of CEC 2017 benchmark with the original method and the proposed type-1 method, which is composed of two inputs and two outputs with 10, 30, 50 and 100 dimensions, obtaining evidence in some functions compared to the original method. Finally, in this chapter experiments were carried out with four benchmark control problems type Mamdani with the original method, type-1 and type-2 by intervals with noise and without noise, as well as two problems of Sugeno control with which experiments were performed with the original method, type-1, type-2 by intervals and type-2 generalized with noise and without noise. It was observed that when applying noise in the controllers, better results are achieved.

We can observe from the results of the experiment with classical benchmark mathematical functions with different numbers of dimensions that using the proposed method, better results are achieved. The results of the experiment with more complex mathematical functions, such as CEC 2015 and CEC 2017, were shown, where we did not obtain the expected results because they are very complex functions with respect to the results that exist in the literature. Likewise, in the control cases that were tested with the proposed method and a comparison with other methods existing in the literature were presented, where we observe that when using noise, better results are achieved. The optimization of this problem is able to obtain better results than with the other methods and are presented in this book. It can be verified that when using perturbation for the controller results have improved.

For the future work of this research, we envision the following: Implement the proposed methodology in other metaheuristic algorithms. In addition, the methodology can be applied to other control problems. We can also test the method with the optimization of other mathematical functions. The methodology with generalized type-2 fuzzy logic can be applied to other control problems. Finally, we can implement the method in a real control cases using hardware implementations of the fuzzy systems.

Appendix

In this section, all the coding of the algorithm proposed in the final version is shown for the reader.

```
%%%%%%%%%%%%%%%%%%%%%%%%%%%%%%%%%%%%%%%%%%%%%%%%%%%%%%
clear all, clc
function [solution, fbest]=hs_funcion(MaxAttempt)
%global ndim;
disp(' it may take a few minutes...');
% the fuzzy system is called
fis=readfis('FHS6');
%plotfis(fis)%//plot the fis
%plotmf(fis,'input',1)%//plot the first membership function
%plotmf(fis,'output',2)
%plotmf(fis,'output',1)
%gensurf(fis)
tic
runtime =1;
for corridas = 30:runtime
%MaxAttempt=1000; % maximum number of attempts
if nargin<2, MaxAttempt =1000; end

if nargin<1,
```

© The Author(s), under exclusive license to Springer Nature Switzerland AG 2020
F. Valdez et al., *General Type-2 Fuzzy Logic in Dynamic Parameter Adaptation for the Harmony Search Algorithm*, SpringerBriefs in Computational Intelligence,
https://doi.org/10.1007/978-3-030-43950-7

```
% CONTROL PARAMETERS %
% functions    1--> Rosenbrock L = -2.048;  H = 2.048;
%              2--> Ackley    L = -32.768; H = 32.768;
%              3--> Griewank  L = -600;    H = 600;
%              4--> Sphere    L = -5.12;   H = 5.12;
%              5--> Rastrigin L = -5.12;   H = 5.12;
%              6--> schwefel  L = -500;    H = 500;


%funstr = '(x1.^2 + x2.^2)/200 - cos(x1).*cos(x2/sqrt(2)) + 1';%//Rosenbrock
%funstr = 'x1.^2 + x2.^2 - 10*cos(2*pi*x1) - 10*cos(2*pi*x2) + 20';%//Rastrigin
%funstr ='20*(1 - exp(-0.2*sqrt(0.5*(x1.^2 + x2.^2))))-
exp(0.5*(cos(2*pi*x1)+cos(2*pi*x2))) + exp(1)';%//Ackley -30 30
%funstr ='(x1.^2 + x2.^2)/200 - cos(x1).*cos(x2/sqrt(2)) + 1'; %funcion griewank -600 600
%funstr ='sum(-x.*sin(sqrt(abs(x))),2)' ; %funcion Schweffel -500 500


end


% convert to a linear function
%f=vectorize(inline(funstr));
ndim=300; % number of independent variables

 for i=1:ndim,
     range(i,:)=[-10 10];
end
%adjust the tone
pa_range=200*ones(1,ndim);



%parameter initialization
HS_size=50; % vector size for solution
HMR=0.1; %harmony memory accepting
PArate=0.75; %pitch adjustment
```

```
% generate the initial solution vector
for i=1:HS_size,
   for j=1:ndim,
      x(j)=range(j,1)+(range(j,2)-range(j,1))*rand;
   end
   HM(i, :) = x;
   %HMbest(i) = f(x(1), x(2));
   HMbest(i) = f(x);
end%%for i
% initialize Harmony Search
for count = 1:MaxAttempt,
      nor=count/MaxAttempt;
         HMR=evalfis(nor,fis);
      %PArate=evalfis(nor,fis);
      %out=evalfis(nor,fis);
      %PArate=out(2);
      %HMR=out(1);
   for j = 1:ndim,

      if(rand >= HMR)
         % new search randomly
         x(j)=range(j,1)+(range(j,2)-range(j,1))*rand;

      else
         %% harmony memory accepting rate
         x(j)= HM(fix(HS_size*rand)+1,j);
         if (rand <=PArate)

            % adjustment tone in given vector
            pa=(range(j,2)-range(j,1))/pa_range(j);
            x(j)=x(j)+pa*(rand-0.5);%0.5
         end
      end
```

```
    end%%for j

    fbest=f(x);

% find the best solution in the vector HS
    HSmaxNum=1; HSminNum=1;
    HSmax=HMbest(1); HSmin=HMbest(1);
        for i = 2:HS_size,
          if HMbest(i) > HSmax,
          HSmaxNum =i;
          HSmax=HMbest(i);
            end
          if HMbest(i)<HSmin,
            HSminNum=i;
            HSmin=HMbest(i);
          end
        end
    % update the current solution if it is better
    if fbest < HSmax,
       HM(HSmaxNum, : ) =x;
       HMbest(HSmaxNum)= fbest;
    end
    solution=x; % save the solution
  %disp(strcat(' Iteracion = ',num2str(count),' Mejor = ',num2str(fbest),'  Minimo =
  ',num2str(HSmin),' Tiempo = ',num2str(toc), strcat(' HMR = ',num2str(HMR), strcat('
  PArate = ',num2str(PArate)))));
  end
  disp(HSmin);
  %save Resultadossumsqaure.mat
  end
  toc
  end
```

%%%
%%%%%%%%%%%%%%%

Type-1 Fuzzy system 1 input and 1 output

[System]
Name='FHS1'
Type='mamdani'
Version=2.0
NumInputs=1
NumOutputs=1
NumRules=3
AndMethod='min'
OrMethod='max'
ImpMethod='min'
AggMethod='max'
DefuzzMethod='centroid'

[Input1]
Name='Iteration'
Range=[0 1]
NumMFs=3
MF1='Low':'trimf',[-0.5 0 0.5]
MF2='Medium':'trimf',[0 0.5 1]
MF3='High':'trimf',[0.5 1 1.5]

[Output1]
Name='HMR'
Range=[0.7 1]
NumMFs=3
MF1='Low':'trimf',[0.55 0.7 0.85]
MF2='Medium':'trimf',[0.7 0.85 1]
MF3='High':'trimf',[0.85 1 1.15]

[Rules]

1, 1 (1) : 1

2, 2 (1) : 1

3, 3 (1) : 1

%%

%Type-1 Fuzzy system 2 input and 2 output

[System]

Name='FHScompleto'

Type='mamdani'

Version=2.0

NumInputs=2

NumOutputs=2

NumRules=9

AndMethod='min'

OrMethod='max'

ImpMethod='min'

AggMethod='max'

DefuzzMethod='centroid'

[Input1]

Name='Iteraciones'

Range=[0 1]

NumMFs=3

MF1='Bajo':'trimf',[-0.5 0 0.5]

MF2='Medio':'trimf',[0 0.5 1]

MF3='Alto':'trimf',[0.5 1 1.5]

[Input2]

Name='D'

Range=[0 1]

NumMFs=3

MF1='Bajo':'trimf',[-0.5 0 0.5]

MF2='Medio':'trimf',[0 0.5 1]

MF3='Alto':'trimf',[0.5 1 1.5]

[Output1]
Name='HMR'
Range=[0 1]
NumMFs=5
MF1='Bajo':'trimf',[0.0053 0.17 0.356]
MF2='Medio':'trimf',[0.34 0.5025 0.69]
MF3='Alto':'trimf',[0.6601 0.851 1]
MF4='Medio_bajo':'trimf',[0.17 0.34 0.52]
MF5='Medio_Alto':'trimf',[0.5 0.67 0.85]

[Output2]
Name='PArate'
Range=[0 1]
NumMFs=5
MF1='Bajo':'trimf',[0.0053 0.17 0.356]
MF2='Medio':'trimf',[0.34 0.5025 0.69]
MF3='Alto':'trimf',[0.6601 0.851 1]
MF4='Medio_Bajo':'trimf',[0.17 0.34 0.52]
MF5='Medio_Alto':'trimf',[0.5 0.67 0.85]

[Rules]
1 1, 3 1 (1) : 1
1 2, 4 5 (1) : 1
1 3, 2 2 (1) : 1
2 1, 2 4 (1) : 1
2 2, 2 2 (1) : 1
2 3, 2 3 (1) : 1
3 1, 2 3 (1) : 1
3 2, 5 4 (1) : 1
3 3, 3 3 (1) : 1

%%

%%%%%%%%%%%%%%

Interval Type-2 Fuzzy system

[System]

Name='FHShmrtype2'

Type='mamdani'

Version=2.0

NumInputs=1

NumOutputs=1

NumRules=3

AndMethod='min'

OrMethod='max'

ImpMethod='min'

AggMethod='max'

DefuzzMethod='centroid'

[Input1]

Name='Iterations'

Range=[0 1]

NumMFs=3

MF1='Low':'itritype2',[-0.583 -0.0833 0.417 -0.417 0.0833 0.583]

MF2='Medium':'itritype2',[-0.08333 0.4167 0.9167 0.08333 0.5833 1.083]

MF3='High':'itritype2',[0.4167 0.9167 1.417 0.5833 1.083 1.583]

[Output1]

Name='HMR'

Range=[0 1]

NumMFs=3

MF1='Low':'itritype2',[-0.5859 -0.08598 0.4141 -0.4193 0.08068 0.5807]

MF2='Medium':'itritype2',[-0.08333 0.4167 0.9167 0.08333 0.5833 1.083]

MF3='High':'itritype2',[0.4246 0.954 1.45 0.62 1.12 1.62]

[Rules]

1, 1 (1) : 1

2, 2 (1) : 1

3, 3 (1) : 1

%%%
%%%%%%%%%%%%%

Generalized Type-2 Fuzzy system

clear all

clc

fis = newgfistype2('FHSg','mamdani','singleton','min','max','min','max','centroid');

 fis = addgvartype2(fis,'input','iteration',[0 1]);

fis = addgmftype2(fis,'input',1,'Low','trigausstype2', [-0.5807 -0.08068 0.4193 -0.4141 0.08598 0.5859 0.5]);

fis = addgmftype2(fis,'input',1,'Medium','trigausstype2', [-0.08333 0.4167 0.9167 0.08333 0.5833 1.083 0.5]);

fis = addgmftype2(fis,'input',1,'High','trigausstype2', [0.422 0.922 1.422 0.5886 1.088 1.588 0.5]);

fis = addgvartype2(fis,'output','HMR',[0 1]);

fis = addgmftype2(fis,'output',1,'Low','trigausstype2',[-0.5833 -0.08333 0.4167 -0.4167 0.08333 0.5833 0.5]);

fis = addgmftype2(fis,'output',1,'Medium','trigausstype2', [-0.08068 0.4193 0.9193 0.08598 0.5859 1.086 0.5]);

fis = addgmftype2(fis,'output',1,'High','trigausstype2', [0.4167 0.9167 1.417 0.5833 1.083 1.583 0.5]);

ruleList=[1, 1 1 1;

        2, 2 1 1;

        3, 3 1 1];

fis=addgruletype2(fis,ruleList);

writegfistype2(fis,'FHSgt.fis'); %guarda el fis

Index

B
Ball and beam controller, 54, 56, 60, 61, 64, 65
Benchmark control problems, 40, 70
Benchmark mathematical function, 2, 13, 24, 32, 69, 70

C
CEC 2015, 24, 69, 70
CEC 2017, 32, 33, 70
Classic Benchmark mathematical functions, 14

D
D.C. Motor Speed Controller, 45, 53
Diversity, 6, 32
Dynamic parameter adaptation, 18, 46, 69

F
Fuzzy harmony search algorithm, 9, 11, 26, 29, 46, 60

I
Inverted pendulum controller, 55, 60, 61, 64, 65
Iterations, 2, 9, 10, 13, 16, 20, 26, 48, 60, 69, 70

M
Mamdani, 9, 40–42, 45, 53, 69, 70

Membership Functions (MFs), 14, 17, 18, 32, 41, 42, 45, 46, 54, 57, 58, 69

N
Noise, 1, 2, 47–53, 60, 61, 63–66, 69, 70

O
Optimization, 1, 2, 5, 20, 21, 45, 65, 69, 70
Original harmony search algorithm, 2, 26, 29, 52, 69

P
Perturbation, 48, 60, 62–65, 69

R
Robot mobile controller, 43–45, 48, 49, 52
Root Mean Square Error (RMSE), 47, 48, 60, 61, 65, 66

S
Sugeno, 40, 53, 54, 57, 58, 69, 70

T
Temperature controller, 42, 43, 48, 49, 52, 53

W
Water tank controller, 40, 41, 48, 49, 52, 53

Printed in the United States
By Bookmasters